LIVING WITH SCIENCE

Fibres

A   S

25p

01A8915

KENT COUNTY LIBRARY

KENT COUNTY LIBRARY
PROJECT LOAN
COLLECTION

WITHDRAWN

Books should be returned or renewed by the last date stamped above.

Melbourne   Sydney
C151165385

Published by the Press Syndicate of the University of Cambridge
The Pitt Building, Trumpington Street, Cambridge CB2 1RP
32 East 57th Street, New York, NY 10022, USA
10 Stamford Road, Oakleigh, Melbourne 3166, Australia

© Cambridge University Press 1985

First published 1985
Reprinted 1988

Printed in Great Britain by David Green Printers Ltd,
Kettering, Northamptonshire.

British Library cataloguing in publication data

Sears, J.
  Fibres and fabrics.—(Living with science)
  I. Fibres
  I. Title   II. Series
  677   TS1540

ISBN 0 521 28581 X

SE

Acknowledgements

The publishers would like to thank the following for permission to reproduce photographs:
Shirley Institute, 4; Farmers Weekly, 10, 12; International Institute for Cotton, 14; Lewis Textile Museum, Blackburn, 24 (bottom); Nigel Luckhurst, and The Scotch House, 24 (top); Nigel Luckhurst 28 (top), 32 (top), 34, 35, 36, 40 (both), 42; Dylon, 28 (bottom); Courtaulds Ltd, 30, 32 (bottom); Douglas Dickins, 33; Topham, 38; Trustees of the British Museum (Natural History), 42 (top).

Illustrations by Colin King
Cover design by Andrew Bonnett

Contents

Unit

1	Threads everywhere	4
2	Spinning a yarn	6
3	Yarns	8
4	Wool	10
5	The golden fleece	12
6	King Cotton	14
7	Smooth as silk	16
8	New fibres from old	18
9	Chemical fibres	20
10	Weaving	22
11	Knitting	24
12	Finishing	26
13	Colour everywhere	28
14	Dyeing	30
15	Printing a pattern	32
16	Keeping fabrics clean	34
17	Washday problems	36
18	Stains	38
19	Choosing the right fabric	40
20	Strange fibres	42
	Index	44

Activities marked with an asterisk use potentially hazardous materials and should only be performed with the supervision of a teacher.

Unit 1 Threads everywhere

Ever since people started to live in cold regions they have needed clothes. At first these were simply to keep us warm and dry. They protected us from the weather. These clothes were made from animal skins. As the number of people got larger it became impossible to kill enough animals to clothe everyone and so people had to find other materials to make clothes from. Also, as society became more complicated, materials were needed for other things than just clothes. Think of all the things that are made from fabrics these days: tents, towels, parachutes, carpets, cushions, curtains, sails; the list goes on and on. Also clothes are not just used for protection any more; many of the things we wear are to make us look nice.

When people started to make fabrics they needed to use materials that were easy to get hold of. At first fabrics were made from **natural fibres**. A fibre is a long thread. Some plants and animals have fibres which can be used to make fabrics. Cotton and flax plants have fibres made from **cellulose**. This is a large sugar molecule or carbohydrate. In cotton the threads are in the seed pod. In flax they are in the

Microscopic views of
cotton
silk
nylon

Microscopic views of

cotton

silk

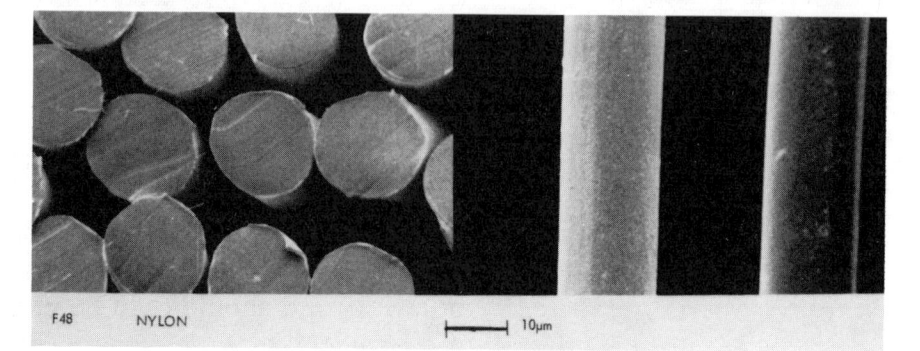

nylon

stem. Flax is used to make linen. Several other plants also give us fibres. These include jute, hemp and sisal. Even the hairs on a coconut can be made into sacks (though I doubt if you would want to wear them!).

The most important **animal fibres** come from the hairy animals or **mammals**. Sheep and goats are particularly good for giving us fibres. Their wool coats can be cut regularly without killing the animal. The animal can also be eaten or can give us milk and cheese. So keeping these animals feeds and clothes us. Some wools are very fine, like the **mohair** that comes from the Angora goat.

The finest animal fibre is silk. This comes from the cocoon of the silkworm. The silkworm is the larva of a moth, just like the maggot is the larva of a house-fly. The silkworm feeds when it is a worm-like larva and then forms a cocoon around itself. In the cocoon it changes into a moth. The moth mates and then lays eggs which turn into silkworms.

Today there are not enough natural fibres to do all the things we want. Man has invented ways of making fibres. Some are made by breaking down natural fibres and then remaking them. Others are made from simple chemicals. These are called **synthetic fibres**. The best-known synthetic fibre is probably **nylon**.

All fibres are different. Some are stronger than others, some are warmer. Man-made fibres tend to be smoother than natural ones but not as warm.

Activity: What are fibres like?

Fabrics are made from fibres. Collect some different fabrics. Make sure you know what each is made from. Unravel lengths of fibre from the fabrics and look at them with a magnifying glass.

Try to answer the following questions:
1 Is the fibre natural or man-made?
2 What does the fabric look like?
3 What do the fibres look like?
4 What differences are there between natural and man-made fibres?

Try wetting some of the fibres and look at them again. Compare the man-made and natural fibres. Compare the wet fibres with the **same** fibres dry. Make drawings of what you have seen.

Questions

1 (a) What is a fibre?
  (b) What is the difference between natural and synthetic fibres?
2 Make a list of the uses of fabrics. For each use try to find out which sort of fibres are usually used.
3 Keeping animals for fibres also has other advantages. List as many as you can. Remember that camels, goats, sheep and llamas are all used for their hair.
4 Copy and complete these sentences:
  (a) The two kinds of natural fibres come from ___ and ___.
  (b) Two kinds of man-made fibres are ___ and ___.
  (c) Cellulose is a large sugar ___.
5 Devise an experiment to test if a fibre stretches more easily when it is wet. Draw a sketch of the apparatus you would use and say how you would carry out the test.

Unit 2 Spinning a yarn

Most natural fibres are too short to be used as they are. They would break easily and not be long enough to weave into fabrics. So fibres have to be turned into a stronger thread or **yarn**. This process of making yarn is called **spinning**. The short fibres are first prepared and combed. The combed fibres are pulled into **slivers**. These are long, loose ropes, with all the fibres running the same way.

Originally these slivers were spun by hand. The spinner held the slivers and twisted them between the fingers of one hand. The twisted sliver was then attached to a spindle. This was held up in the other hand. The spindle was made to spin in the opposite direction from the twisting. The twisted slivers were thus wound into a tight yarn. As this happened the spindle slowly went down until it touched the ground. The spinner then stopped and wound the yarn onto the bobbin of the spindle. Then the next piece of yarn was wound and so on. As you might imagine this took a very long time!

The invention of the spinning wheel speeded things up. At first it was hand-worked and the twisting and winding had to be done separately, but soon an extra piece, the **flier**, was invented. This fitted over the bobbin. As the wheel turned the flier twisted the thread and wound it onto the bobbin at the same time. The spinner still fed the slivers into the flier, but of course this could be done much faster than simple hand spinning. Even so, it took five or six spinners to make enough yarn for one weaver.

So people looked for ways of speeding up spinning. One of the first machines to do this was the **Spinning Jenny**. This was invented by James Hargreaves in 1768. The machine had a moving carriage. It pulled the thread out as it moved one way and wound the thread onto bobbins as it moved the other. This allowed one spinner to work as many as eighty spindles at once.

A further improvement came from Richard Arkwright in 1769. His **water frame machine** had rollers for collecting several slivers together into a thicker strand or **roving**. The roving was then twisted into a stronger and more regular yarn.

Arkwright's rollers and Hargreaves' moving carriage were combined in Samuel Crompton's **Spinning Mule**.

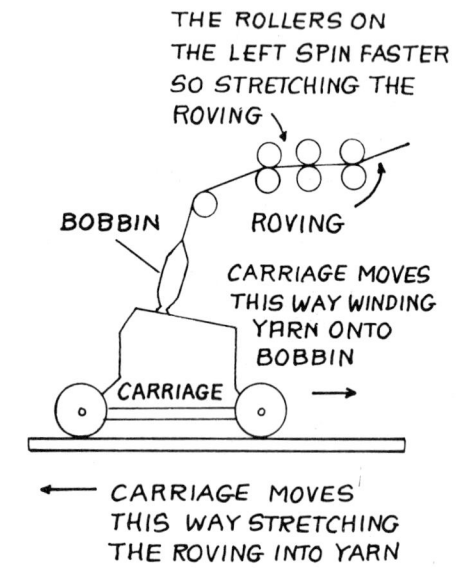

A SPINNING MULE

Today, most spinning is done on **ring frame machines**. The roving is fed into a ring which spins round and up and down the bobbin. The spinning of the ring twists the thread. The spinning also winds the thread onto the bobbin. The up and down movement winds the thread evenly onto the bobbin.

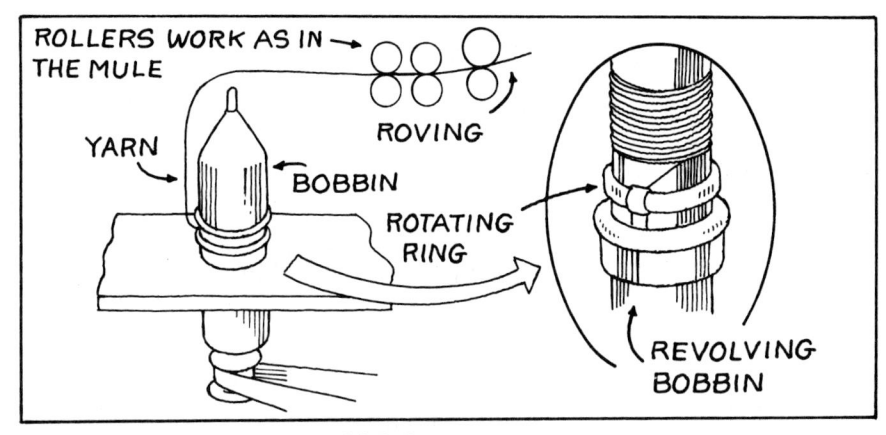

A RING FRAME MACHINE

Activity: Make your own spindle

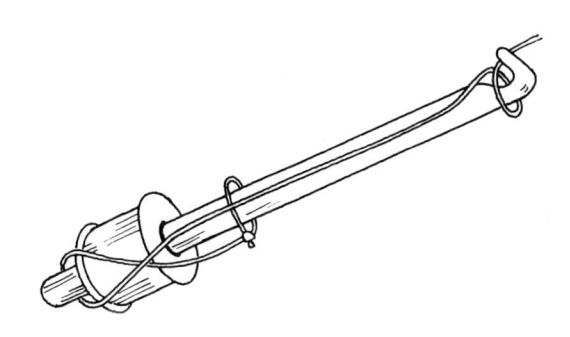

You can make a hand spindle out of a cotton reel and a piece of dowel. The end of the dowel needs to be carved into a hook. The other end is glued in place into the cotton reel. The dowel should be about 30 cm long. Get some staple fibre (short fibres) and spin about half a metre of yarn. You do this by pulling the fibres from the sample and twisting them with your fingers. Wrap the yarn around the spindle as shown in the diagram. Hold the staple fibre end of the yarn in one hand. Spin the spindle with the other until the yarn is well twisted. Pull out more staple and let the twist run into the drawn out fibre. Do this until the yarn is too long to cope with. Unhook it from the spindle and wind it onto the spindle shaft (cotton reel). Fix it again and repeat as before. How easy did you find spinning in this way?

All spinning was done this way until about AD 1500.

Questions

1. What is a yarn?
2. Why do fibres need to be spun into yarns?
3. What is a flier and why is it needed?
4. Copy and complete these sentences: Short _____ from a plant or animal are combed into _____. These are then twisted together. Several slivers are drawn out between rollers to make a _____. This is finally spun into a _____.
5. Copy the drawing of the mule and explain how it works.
6. Spinning needed to get faster to make enough yarn for the weavers. What invention of John Kay made this even more necessary? (You may need to look this up in the library.)

Unit 3 Yarns

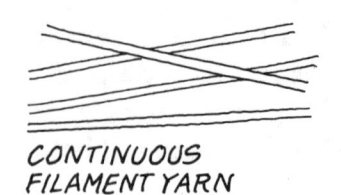

CONTINUOUS FILAMENT YARN

STAPLE YARN

LOOSE TWIST

MEDIUM TWIST

HIGH TWIST

Yarns are made by spinning the fibres together. There are two main sorts of yarn. They are made from either staple fibres or continuous filaments. A **staple fibre yarn** is one made from short fibres. These need twisting and spinning. A **continuous filament yarn** is made from very long single fibres. Man-made fibres and silk have very long single fibres. These do not need to be spun together like staple fibres, but they do need twisting. Some man-made fibres are cut up into short lengths and then spun like wool or cotton. This gives the yarn a different texture (feel).

Filament yarns are smooth and often shine. Staple yarns are fuzzy with hairs sticking out of them. They are also heavier than filament yarns. So, if you wanted something warm to wear, you would make it out of a staple yarn. If you wanted something shiny and pretty you would be more likely to use a filament yarn.

Yarns come in different thicknesses. The thickness of a filament yarn is measured in **denier**. Actually, the denier tells you how heavy a given length of yarn is; so the thicker the yarn the higher the denier number. Nylon stockings often tell you what denier yarn is used. A 30 denier stocking will be thicker than a 15 denier stocking.

Yarns also have different amounts of twist in them. A loose twist will give a soft, weak fabric, but one that is warm. A medium twist gives a stronger yarn which is often used as the support thread in weaving (the warp). Very high twisting is used in yarns like sewing cotton where very hard wear is needed.

SINGLE YARN

TWO-FOLD YARN MADE FROM TWO SINGLE YARNS

THREE-FOLD YARN MADE FROM THREE SINGLE YARNS

CABLED OR TWO-PLY YARN MADE FROM TWO LOTS OF TWO-FOLD YARN

Some yarns are made by winding several single yarns together. The extra threads make the yarn stronger. If you have ever bought wool you will have heard the expressions 'two-ply' or 'three-ply'. The **ply** tells you how many folded yarns are twisted together. So two-ply is two folded yarns. By using two-ply yarns it is possible to have a loose twist with great strength.

Finally, the arrangement of fibres in a yarn can be important. This only really happens in staple yarns. Compare a really good linen handkerchief with a tea towel (make sure both are linen). The tea towel is much coarser and less shiny. This is because the fibres in its yarn are shorter and arranged unevenly.

Activity: Comparing different yarns

Get some sewing cotton, several lengths of wool (different plys), some worsted yarn (or fabric that you can cut up), some silk, some handkerchiefs (old linen ones), and an old tea towel. Unravel some of the yarn from the fabric so you have threads of all the different things listed.

Use a microscope or hand-lens to look at the yarns. How many strands are there in each yarn? How are they wound together? Are any of them hairy? How many twists are there in a centimetre?

Unravel the threads. Which ones are staple yarns? Which ones are filament yarns?

Try separating the fibres. Are the fibres arranged parallel in any of the yarns?

Use equal lengths of the different sorts of yarn. Hang the lengths up and attach weights to them. Which yarns are strongest?

You may find it particularly interesting to compare the different types of wool.

Fill in your results in a table as shown.

SOURCE OF YARN	
DIAGRAM OF YARN	
TWISTS PER CM	
FILAMENTS OR STAPLE	
WEIGHT THAT JUST BREAKS YARN	

Questions

1 Make a table to show the differences between staple yarns and filament yarns. Draw it as shown below.

	STAPLE YARNS	FILAMENT YARNS
FIBRE LENGTH		
THICKNESS		
SMOOTHNESS		

2 How is the thickness of filament fibres measured? If you can, find out exactly what the numbers mean.
3 Make a careful drawing of a two-ply yarn. Why are different sorts of woollen yarn made?
4 From the figures given below draw a graph of fibre production since 1900.

Fibre production

Year	Man-made	Wool	Cotton	World population (millions)
	(thousands of tonnes)			
1900	1	730	3 162	1 550
1930	208	1 002	5 927	2 000
1935	490	980	6 055	
1940	1 132	1 134	6 970	2 200
1945	618	1 034	4 667	
1950	1 681	1 057	6 647	2 500
1955	2 545	1 265	9 492	2 700
1960	3 310	1 463	10 112	3 000
1965	5 390	1 493	11 604	3 300
1970	8 132	1 602	11 686	3 632
1974	11 018	1 502	13 787	3 890

(a) Why do you think man-made fibre production went down in 1940–45?
(b) Why did wool production stay the same at this time?
(c) Has total fibre production increased? If so why?

Unit 4 **Wool**

It is known that people were wearing woollen garments at least 10 000 years ago. One famous ancient area is called 'The Land of Wool' – this is what Babylonia means. It was called this because of the number of sheep found there.

In Britain sheep have been kept for as long as we have kept records. About 3000 years ago the ancient Phoenicians came to Britain from the Mediterranean. They traded metal goods for woollen ones. In ancient Rome one writer said that British wool was 'finer than a spider's web'. When the Romans conquered Britain they set up a wool trade. Many Roman sheep farmers lived in villas outside the towns. You can still see the remains of these today. The Saxon invasion of Britain caused the collapse of the wool trade. But within two hundred years we were again sending our wool to Europe.

After the Norman conquest the British wool trade grew to be very important. Merchants became rich by selling wool to clothmakers abroad. The King gained much of his money from taxes on wool. The trade was at its biggest in about 1320. Many inns and pubs called the Woolpack can be seen in the country. Many of these were where the 'broggers', or woolmen, stopped when carrying their wool to market.

Gradually the wool trade was replaced by the cloth trade. Instead of selling the wool to be made into cloth abroad, the British clothmakers used it and sold the finished articles abroad. The clothmakers were formed into guilds that met in a Guildhall. Many towns still have Guildhalls. To become a master clothmaker you had to produce a piece of cloth that was approved by other masters as being good quality. This was called the **masterpiece**. The clothmakers learnt their trade from the Flemish weavers who were encouraged to come to England by Edward III. Until then only a few of our cloths, like Lincoln Green, had been as good as foreign ones.

Since that time the story has been one of mechanization. Yorkshire became the industrial centre of the wool weaving industry. Wool has been such an important part of our heritage that there are many fascinating stories to be told about it. Queen Elizabeth I passed a law saying that everyone must wear a woollen cap out-of-doors. Charles II said that everyone was to be buried in a woollen shroud. The Lord Chancellor still sits on a woolsack to remind him that much of the country's wealth comes from the wool trade.

Activity: British sheep

There are over thirty different breeds of sheep kept in Britain. Most of them produce coarse wool that is used in carpet making or for blankets and furnishings. The finest wools are produced abroad and these are used for clothes. There are some British breeds that produce wool for garments, e.g. Shetland wool. Get a large outline map of the British Isles. Using the information below, mark in the areas where sheep are mainly found. Also mark in the main wool towns. Use different signs for the different sorts of cloth made.

The main towns are Axminster, Wilton, Witney, Carmarthen, Cardigan, Kidderminster, Leicester, Bradford, Huddersfield, Hawick, Edinburgh, Glasgow and Perth.

You may get pictures to stick on your map and more information from the Wool Marketing Board.

Area	Breed of sheep
CORNWALL	South Devon, Dartmoor, Whiteface Dartmoor
DEVON	Devon Longwool, Devon Closewool, Exmoor Horn
DORSET	Dorset Down, Dorset Horn
HAMPSHIRE	Southdown, Hampshire Down
KENT	Romney
OXFORD	Oxford Down
SOUTH WALES	Welsh Mountain Mountain, Black Welsh
WORCESTERSHIRE	Radnor, Ryeland
NORTH AND CENTRAL WALES	Clun Forest, Kerry Hill
EAST ANGLIA	Suffolk
SHROPSHIRE	Shropshire
DERBYSHIRE	Derbyshire Gritstone, Lonk
LINCOLNSHIRE	Lincoln Longwool, Leicester
LAKE DISTRICT	Herdwick, Rough Fell
PENNINES	Swaledale
YORKSHIRE	Wensleydale, Teeswater, Dalesbred
CHEVIOTS	Cheviot
NORTHUMBRIA	Border Leicester
CENTRAL SCOTLAND	Blackface
NORTHERN SCOTLAND	North Country, Cheviot

Questions

1 Complete this crossword.

Across
1. The Land of Wool.
4. _____ sack is what the Chancellor sits on.
8. What clothmakers belong to.

Down
2. The funny name for a woolman.
3. The 'Woolpack' is one.
5. Where the wool comes from.
6. What wool is made into.
7. Where Roman farmers lived.

2 Write a short essay called 'The history of wool in Britain'.

Unit 5 The golden fleece

Even though man-made fibres are now used more than natural fibres, wool is still very important. There are many breeds of sheep and the length of fibres are different for each.

Wool is a special sort of hair and is cut (shorn) from a sheep once a year. The shearers take off all the wool at one time. The cut wool is called a **fleece**.

The fleece is first washed with soap to get rid of oil and sweat. This is called **scouring**. The fleece is then treated with acids and heat. This gets rid of any straw or burrs that were caught in the fleece. This process is called **carbonizing**. The different sorts of wool from the fleece are then sorted and the fibres are smoothed out by a **combing process**. The wool is then **spun** into yarn. Some wool, though, gets an extra combing until all the fibres are nearly parallel. This produces a yarn called **worsted**. The worsted yarn is smoother and less hairy. It is used in making suits.

Sheep shearing

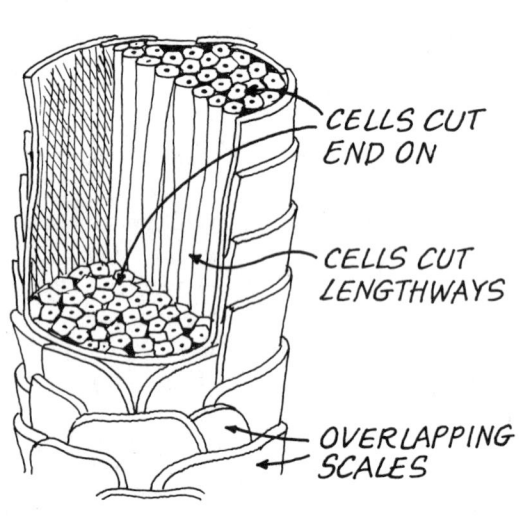

A WOOL FIBRE CUT OPEN
- CELLS CUT END ON
- CELLS CUT LENGTHWAYS
- OVERLAPPING SCALES

Wool yarn is very special in the way it behaves. If you look at it under a microscope you will see that it is scaley. The scales overlap and run towards the tip of the wool. Inside, the wool is made of fine thin cells. These run along the length of the fibre. The coarser fibres have a hollow centre. The fibres are also naturally curly. All of these features make wool behave in a special way.

If you stretch the fibres they will spring back to their original size. This is very useful when you want a garment to keep its shape even with a lot of use. It also means that wool fabrics do not crease permanently. Because the fibres have a lot of air in them, wool is very good at insulating. Another feature of wool is that it can take in water and not feel as if it is wet. Up to one third of its weight can be water and the wool will still feel dry. Also, taking in water makes the wool warm and so, if you have to be out in bad weather, wear wool. This is specially useful for people who go walking in mountains. Wool trousers are better than cotton ones because they keep you warm when they are wet. Water also runs off wool very easily so if it rains the drops fall off it.

One problem with wool is that if you rub the fibres hard in warm water the scales all cling together. This makes what we call **felt**. Felt was probably one of the first woollen fabrics. The same thing can happen with knitwear if you wash the wool too roughly. The fibres cling together and the garment shrinks. Once this has happened the fibres cannot be pulled apart, so you must be careful!

Activity: What is wool like?

Take some wool yarn. Hang it from a support. Tie a weight holder to it. Stand a ruler up against it and mark the position of the holder. Now add a small weight so that the yarn stretches. Take the weight off. Does the fibre go back to its original length? Keep adding weights until the yarn stays stretched. How far could you stretch the yarn before you damaged it?

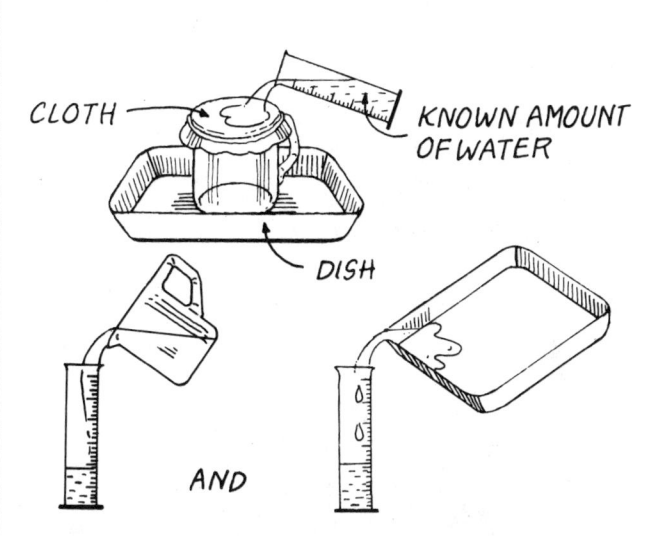

Get some woollen cloth and fix it over the top of a jug with a rubber band. Pour a known amount of water into the cloth. How much water went through into the jug? How much ran off? (You will need to catch this by standing the jug in a dish.) How much has been kept by the wool? Does the wool feel wet?

Get an old jumper. Wash one part in warm water by gently rubbing. Wash another part in hot water and rub it hard. What is the difference at the end? (Make sure it is a jumper you no longer want!)

Write a summary about what these experiments tell you about wool.

Questions

1 Explain the processes of scouring and carbonizing. Why is each done?
2 What is the difference between an ordinary wool yarn and a worsted yarn?
3 Make a list of all the useful things wool can do. For each one try to explain why the fibres do what they do.
4 Try to explain why people in cold places and people in hot places both wear wool clothes.
5 By further reading find out and make a list of the different things that wool is used for.
6 Some wools come from animals other than sheep. Find out what sort of animals give you angora, cashmere, mohair, alpaca and vicuna wools.

Unit 6 King Cotton

Cotton boll

Cotton is one of the fibres that has been used by man for a long time. Cloth made from cotton has been found in ancient graves in Egypt and Peru. Much of this old cloth was of very high quality, even though it was all made by hand. Cotton has probably been used for at least 3000 years.

Cotton is a plant that grows only in countries which have a lot of sun and rain. It needs about six months of hot weather with heavy rainfall. The plant grows into a small bush of about one metre high. The flowers are small and white and only last two days. After the petals have fallen a small green seed pod grows. After six to eight weeks the pods burst open. Inside there are lots of white hairs. The pods look like balls of cotton wool. The pods are called **bolls**.

The bushes are grown in fields in long rows. The bolls are picked at harvest time by hand. In America in the 1800s it was the slaves who worked in the cotton fields. The main slave farms were started when the **cotton gin** was invented. This was a machine which separated the cotton fibres from the seeds and stalks. About 35 000 000 slaves were brought in from Africa, but only 15 000 000 arrived in America alive.

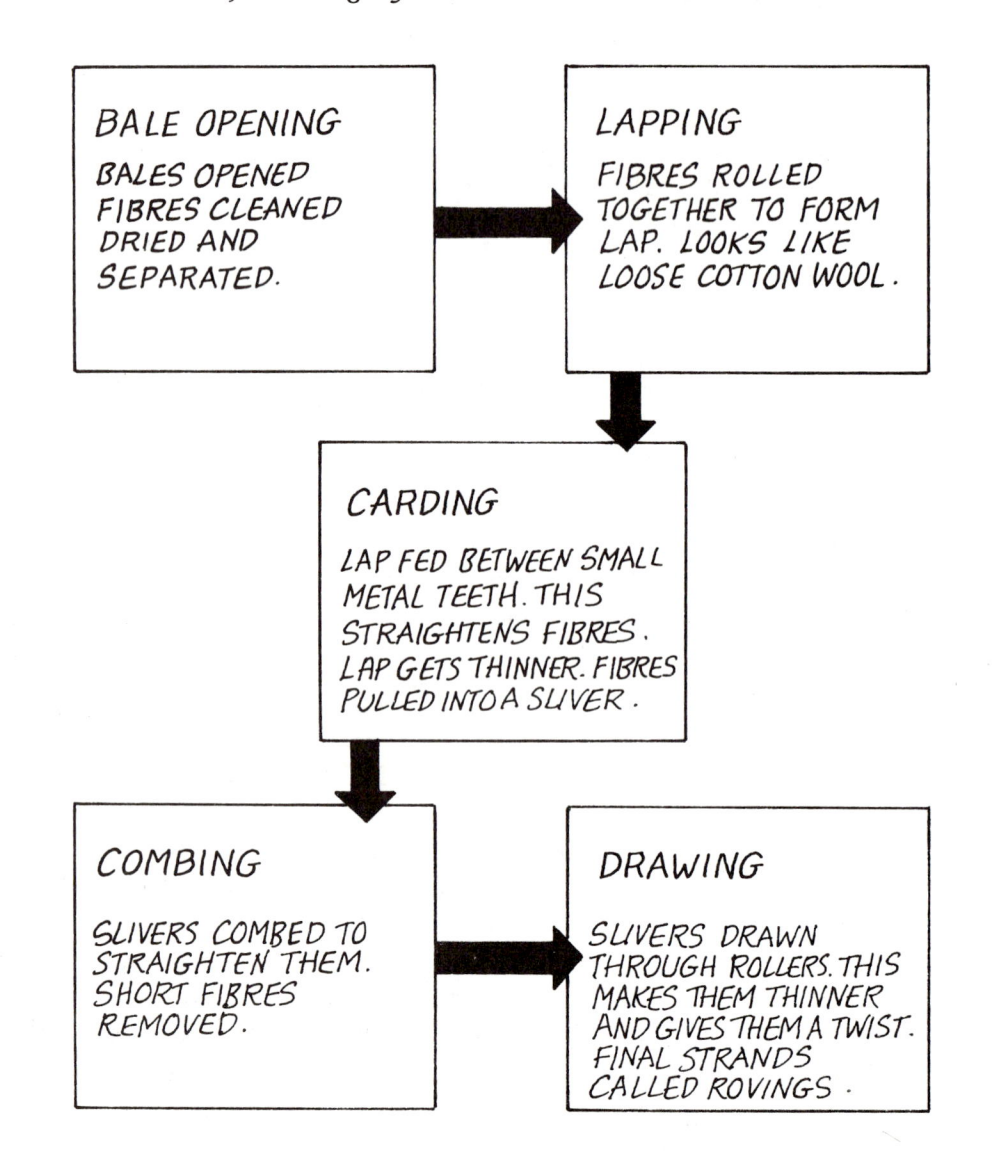

Processing cotton

BALE OPENING
BALES OPENED FIBRES CLEANED DRIED AND SEPARATED.

→

LAPPING
FIBRES ROLLED TOGETHER TO FORM LAP. LOOKS LIKE LOOSE COTTON WOOL.

↓

CARDING
LAP FED BETWEEN SMALL METAL TEETH. THIS STRAIGHTENS FIBRES. LAP GETS THINNER. FIBRES PULLED INTO A SLIVER.

↓

COMBING
SLIVERS COMBED TO STRAIGHTEN THEM. SHORT FIBRES REMOVED.

→

DRAWING
SLIVERS DRAWN THROUGH ROLLERS. THIS MAKES THEM THINNER AND GIVES THEM A TWIST. FINAL STRANDS CALLED ROVINGS.

These evil practices were brought to an end by Abraham Lincoln at the end of the American Civil War.

Most cotton is still picked by hand because less rubbish gets into it this way. The picked bolls are taken to the gin to have the stalks and seeds removed. Each boll produces as many as 20 000 fibres. These can be from 10 to 60 mm long. The longer they are the better the cotton is in the end. The fibres are pressed into bales. These are sent to the factories to be processed. This sort of processing is used for all staple fibres. (See flow diagram.)

Cotton is strong and wears well. It is also fairly cheap. This makes it ideal for dresses, shirts, pyjamas, underclothes, etc. Many different cloths are made from cotton: denim, muslin, sateen, gingham and many others. No wonder cotton is sometimes called the 'King of Fibres'.

Activity: What is cotton like?

Although cotton is very tough it does not keep you very warm. Wrap some cotton cloth twice round a beaker and hold it in place with a couple of rubber bands. Pour a known amount of boiling water into the beaker. Put a thermometer in and take the temperature every minute for ten minutes.

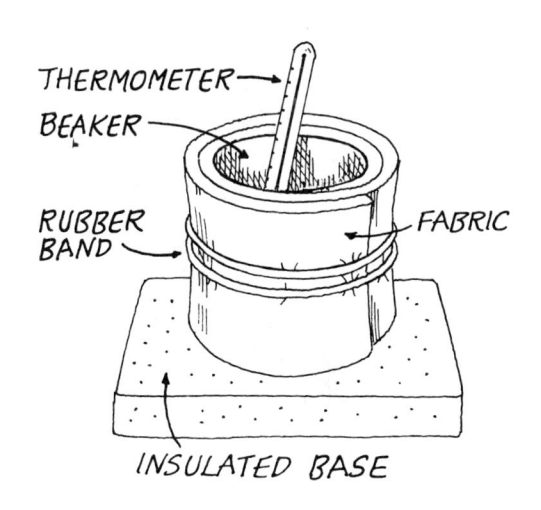

Do the same experiment with other fabrics. Which ones insulate the beaker best? How does cotton compare to the others?

You can also repeat the experiments you did with wool at the end of the last chapter. Does cotton stretch as far? Is it as good at keeping out water? Does it take in water as easily?

Short fibres of cotton are used to make the substance we call cotton wool. You might like to try combing this and then spinning it as described in Unit 2. How easy was it? Now you can see why the longer fibres are used to make yarn!

Questions

1 Cotton is grown mainly in: (a) northern India; (b) eastern China; (c) south and eastern United States; (d) Mexico; (e) eastern Brazil; (f) northern Nigeria. Colour these areas on an outline map of the world.
2 Why are these areas good for cotton growing? Try to find out **exactly** what they have in common.
3 (a) What does a cotton gin do? Why do you think it made the cotton cheaper?
   (b) Why is most cotton picked by hand?
4 Why do you think nature has put hairs in the cotton seed pod?
   (Clue: animals eat seeds, e.g. peas)
5 Copy the flow diagram to show how bales of cotton are processed.
6 Why do you think cotton is sometimes called the 'King of Fibres'?

Unit 7 Smooth as silk

Silk is the most unusual natural fibre. It is the only one which is a long continuous filament. It was first used in China about four thousand years ago. The Chinese guarded the secret of its production for centuries. But eventually other people found out how to grow silk and its manufacture spread to other countries. Nowadays, Japan is the main producer; China, India, France and Mexico also produce it.

Silk is made by the silkworm. The most commonly used silkworm is *Bombyx mori*. This latin name means 'silkworm of the mulberry tree'. The caterpillar only eats mulberry leaves. When it is fully grown the caterpillar starts to change shape into an adult moth. During this stage it is called a **pupa**. To do this it needs to find somewhere sheltered. To help protect itself from being eaten it spins a **cocoon** around itself. The pupa lives inside the cocoon, protected from attack by other insects – but not from man! It is because of its cocoon that we keep the caterpillars. The cocoon is spun from up to a mile of silk. The caterpillar makes the silk in special glands. It forces the silk from the gland and it hardens in the air.

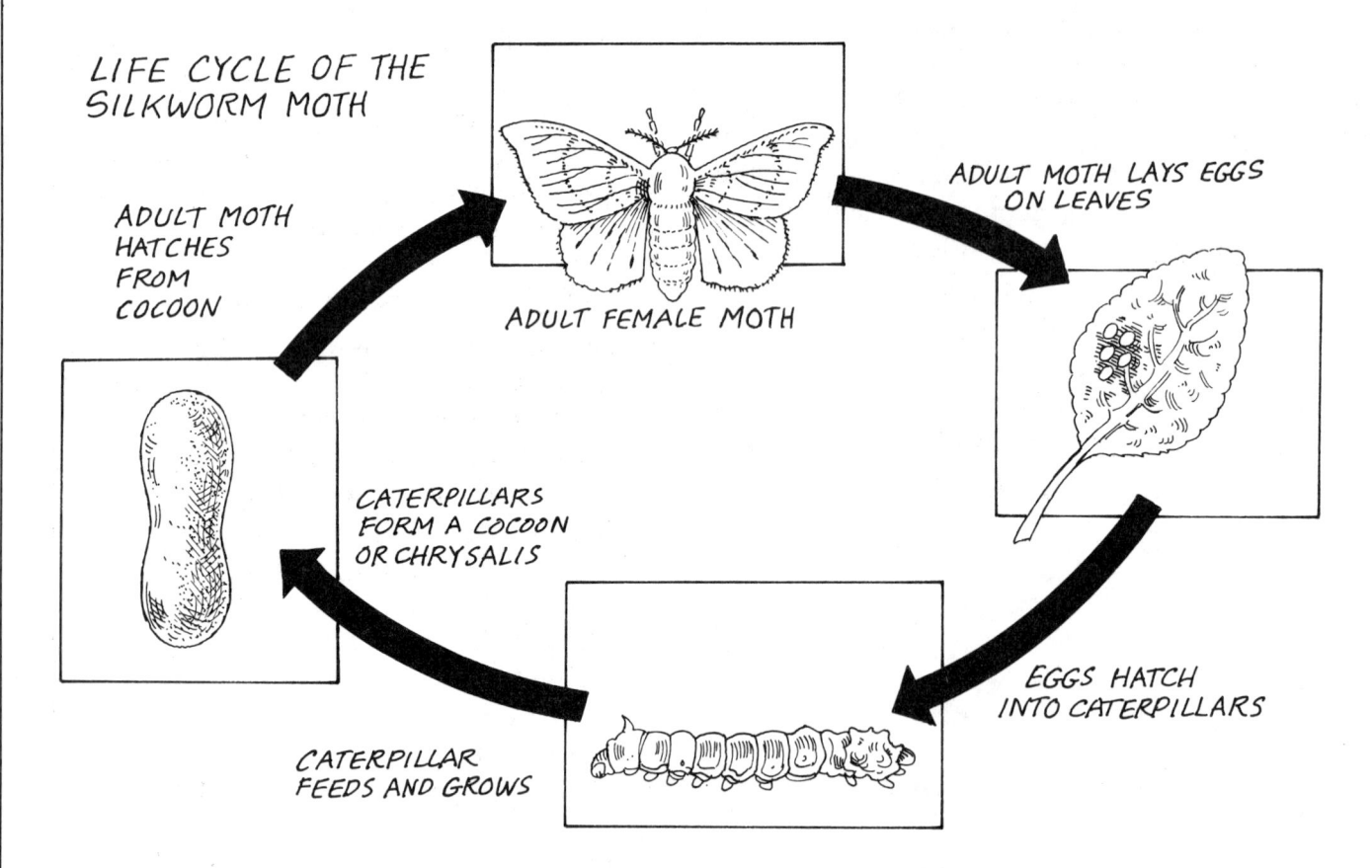

LIFE CYCLE OF THE SILKWORM MOTH

- ADULT FEMALE MOTH
- ADULT MOTH LAYS EGGS ON LEAVES
- EGGS HATCH INTO CATERPILLARS
- CATERPILLAR FEEDS AND GROWS
- CATERPILLARS FORM A COCOON OR CHRYSALIS
- ADULT MOTH HATCHES FROM COCOON

The pupa is killed by heating and then the cocoon is softened in hot water until the silk starts coming loose. The silk threads are then picked up and wound onto a reel. Because the threads are so fine it is normal to wind about five threads into one filament at the same time. A normal cocoon gives about half a mile of unbroken threads. This silk is called **reeled silk**.

The outside of the cocoon is covered in short lengths of silk. These, together with any silk from broken or damaged cocoons, are spun together to form a yarn in the same way that cotton or wool is. This sort of silk is called **spun silk**. It is not as fine or smooth.

The silk then needs to be **degummed**. This gets rid of the sticky gum that the silkworm used to keep the cocoon together. Much of the silk yarn is also **thrown**. This means it is given an artificial twist. This gives it more texture than the pure silk yarn.

As you can imagine it takes a long time to unwind a cocoon. Even though the unwinding is mechanized much of the work of rearing the worms is done by hand. The time and the number of workers involved means that silk is always going to be an expensive fabric. It is warm and has a very attractive appearance. Silk garments are usually used on special occasions.

Activity: Growing silkworms

It is possible to get silkworms from suppliers. If they are bought as eggs you can grow the worms on lettuce leaves rather than mulberry. The worms need a lot of care but you will get instructions with the eggs. They will feed and grow, and moult at regular intervals as they get bigger.

When they have moulted for the final time they can be put into egg boxes to spin their cocoons. This takes about 24 hours but watching them during the process is fascinating. When you have the cocoons it is easy enough to get the silk from them.

Try rubbing the silk from the outside of the cocoons. These are the short fibres. Save these and try hand spinning them into a yarn. Put the rest of the cocoon into the oven at 200 °C for 30 minutes. This kills the pupa. Now heat five or six cocoons in a saucepan of water. When you see the threads coming loose turn off the heat. Pick up a loose end with some tweezers and start to wind it onto a reel. Gradually pick up the other loose ends. They should stick to one another. Keep winding with the cocoons bobbing in the water. If you are only winding by hand you will find that it takes several days to get all the silk off!

You may want some more silkworms later. If so keep some cocoons. These will hatch into moths later.

Questions

1 What makes silk different from other natural fibres?
2 On your world map of cotton production mark the areas of silk production.
3 What are the two sorts of silk? What is the difference between them?
4 Why does the silkworm spin a cocoon?
5 Why is it important for the farmers not to use all the cocoons to make silk?
6 Why is silk such an expensive fabric?
7 Copy and complete these sentences:
Bombyx mori means _____ eater. It is a kind of _____. The larva (caterpillar) makes the silk when it turns into a _____. The animal uses the silk to _____ itself from attack.

Unit 8 New fibres from old

All natural fibres suffer from one major disadvantage – they take a long time to make. Sheep need time to grow, so does cotton, and the time needed to make silk is enormous. Also, the best-looking fabric (silk) costs so much that most people cannot afford it. So it is hardly surprising that people looked for some man-made fibres to act as substitutes.

In 1846, a German called Schonbein developed a process to make a sticky gum from cotton plants. He used the gum to make an explosive material called **gun cotton**. Of course, explosive material was no good for clothes! The big step came when an Englishman, Joseph Swan, invented **artificial silk**. He was trying to make fine threads to act as filaments for the electric light bulb. He invented a way of squirting a sticky gum through an apparatus shaped like a shower head. The fibres were set by being squirted into an acid bath. The fibres he made were no good for what he wanted, so it was left to other people to see their use in fabrics.

Count Chardonnet in France worked on the problem. He developed the Swan process and made the first clothes from artificial silk. At first the clothes were a little explosive, like gun cotton, but he solved this problem. His garments, called miracle garments, were soon replaced by those made from **rayon**.

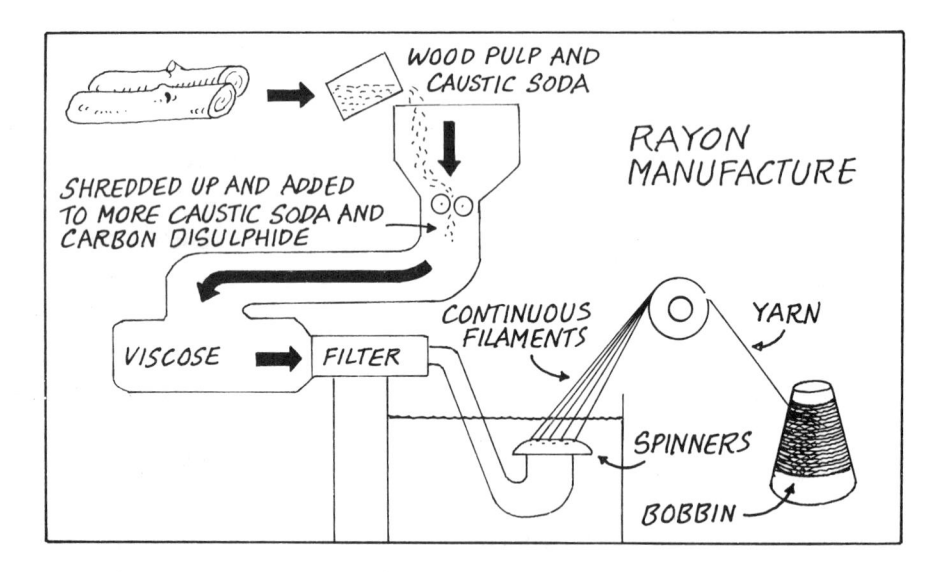

Rayon is made from plants. All plants have **cellulose** in them. This chemical forms long fibres. If cheap plant material is dissolved in caustic soda and other chemicals it forms a sticky gum. The fibres get all mixed up. The gum is then squirted through tiny holes into an acid. The acid hardens the gum into fibres. The fibres are made from cellulose which has been reorganized (**regenerated**). Sometimes it is possible to harden the fibres by squirting them into air.

Today there are many different types of regenerated fibres. Rayon and acetate fibres are two common examples.

They can be spun out in continuous filaments or the filaments can be cut up and turned into short fibres. These are then spun into thread like cotton fibres or wool. Regenerated fibres are cheap. They can be used for a wide variety of garments, especially those expected to have a short life. By varying the yarn they can be used to make cloth or knitwear. Rayon was the first widespread artificial fibre. It had a lustre like silk and could be afforded by nearly everyone.

Activity: Making rayon*

You may be allowed to do this experiment at school. It is possible to make simple rayon fibres from paper towels. Paper is made from trees and so contains cellulose.

Put about 50 cm³ of ammonia solution in a beaker. The stronger the solution the easier it is, but diluted ammonia will give you fibres if you are patient. Add copper carbonate powder to the mixture. Stir all the time. Keep adding the powder until some of it is undissolved. Leave the solution for five minutes.

Tear a paper towel up into tiny pieces while you wait. Pour the solution into a mortar. Add bits of the paper towel and crush them with the pestle. Do this until the liquid goes really sticky and thick. The weaker the ammonia the longer this takes.

Put 100 cm³ of dilute sulphuric acid in another beaker. Suck up the sticky gum into a syringe. Squirt the gum into the acid with the end of the syringe under the surface. You will see small weak fibres forming. These dissolve fairly quickly so look carefully.

Obviously the real manufacturing has to be more complicated to stop the fibres dissolving.

Questions

1 Why did people try to make artificial fibres?
2 Briefly describe how rayon is made. Use diagrams.
3 Why are these fibres called 'regenerated fibres'?
4 Try to find out what sort of fibre is used in making the following fabrics: Acetate; Arnel; Brocade; Dicel; Durafil; Evlan; Faille; Lancola; Tricel; Viscose.

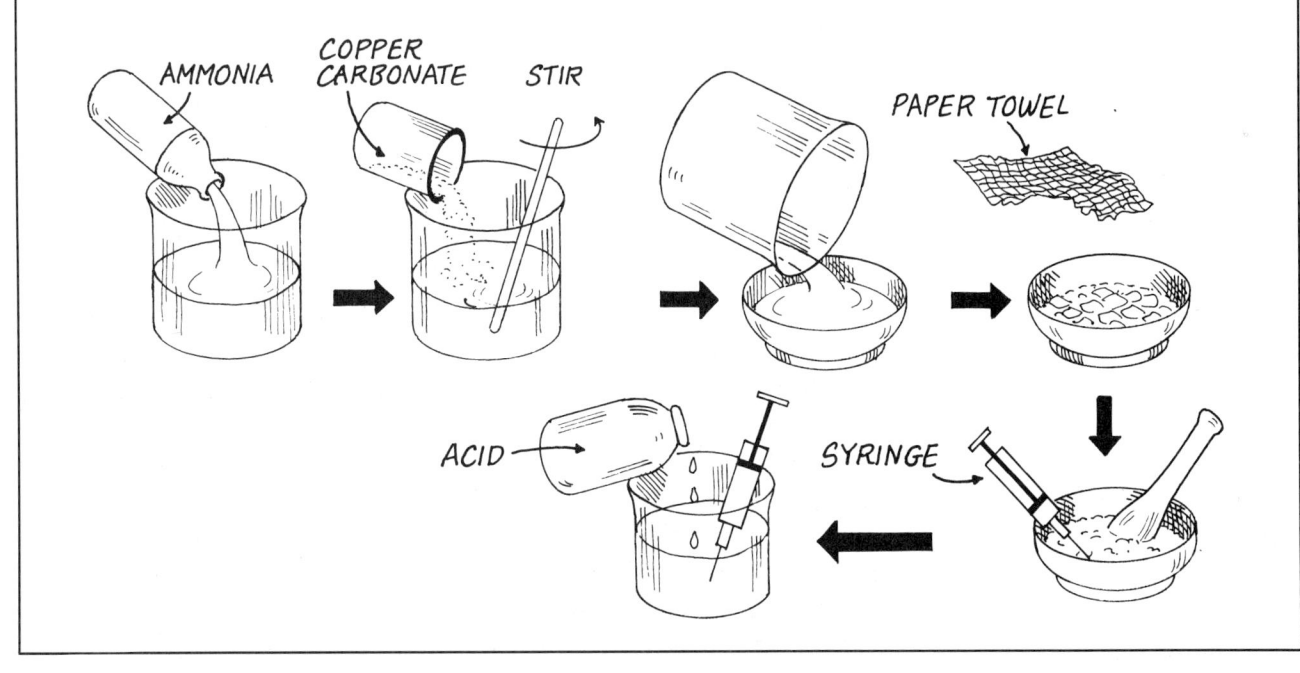

Unit 9 Chemical fibres

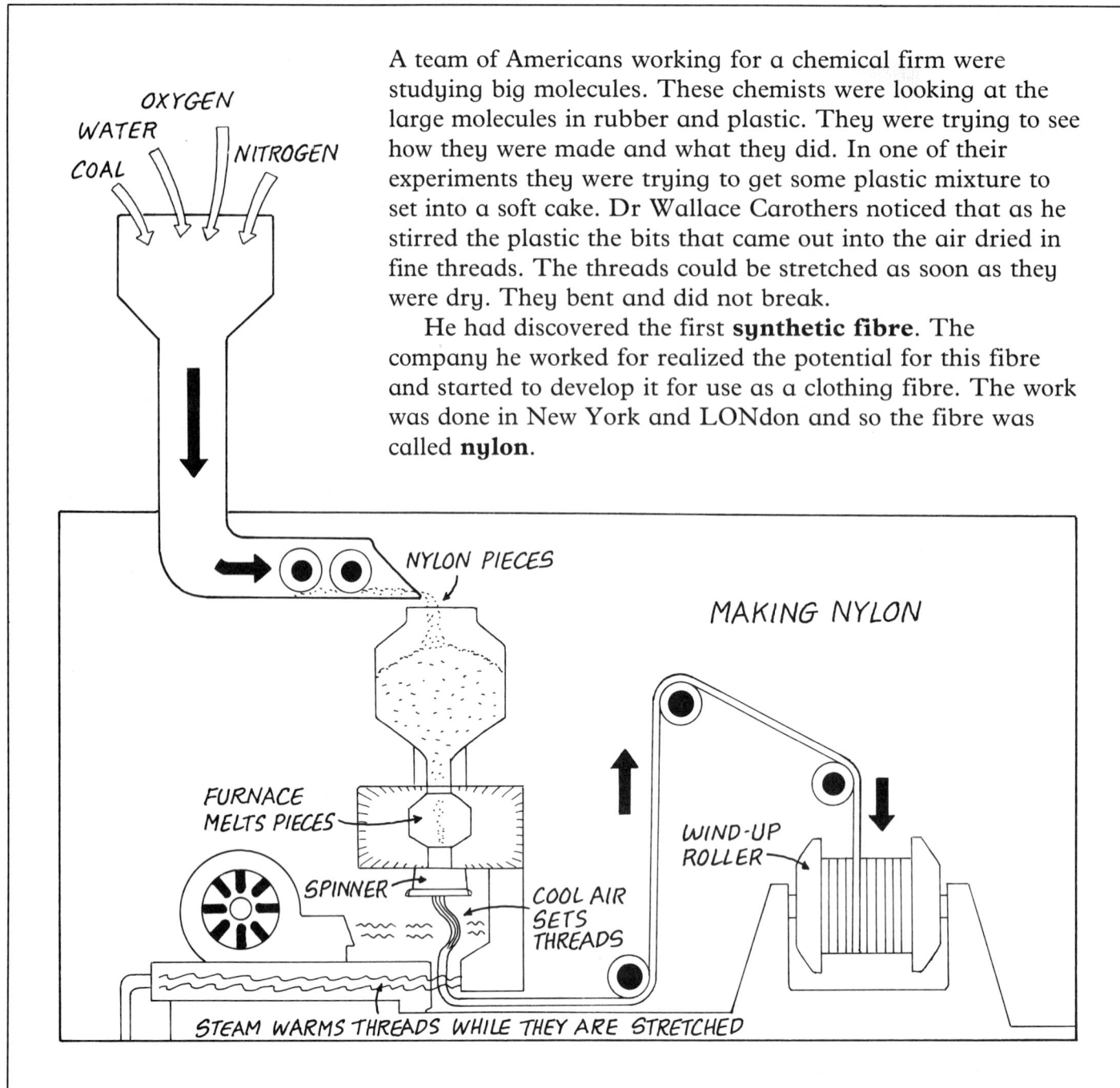

A team of Americans working for a chemical firm were studying big molecules. These chemists were looking at the large molecules in rubber and plastic. They were trying to see how they were made and what they did. In one of their experiments they were trying to get some plastic mixture to set into a soft cake. Dr Wallace Carothers noticed that as he stirred the plastic the bits that came out into the air dried in fine threads. The threads could be stretched as soon as they were dry. They bent and did not break.

He had discovered the first **synthetic fibre**. The company he worked for realized the potential for this fibre and started to develop it for use as a clothing fibre. The work was done in New York and LONdon and so the fibre was called **nylon**.

Nylon was first used in the war to make parachutes. After the war it was rapidly used for all kinds of clothes. The great thing about nylon is that it is tough and cheap. It has a good finish and does not need a lot of drying. Indeed it dries so easily that it is called a drip-dry fibre. Like the regenerated fibres it can be spun in a continuous filament or cut up and spun into a staple yarn.

There are many different sorts of synthetic fibres now. They can be divided into three main groups. The polyamides (or nylons), the polyesters (Terylene is the best example) and the acrylic group (Courtelle is a good example). All these fibres are made in the same sort of way. Raw materials are mixed together to make a sticky plastic liquid. The liquid is forced through holes so that fine threads come out. The threads are dried in the air and then wound onto rollers.

Synthetic fibres all have the same basic features. They are hard wearing. They dry easily and wash well. They resist attack by chemicals and pests. They have a good finish when woven. They hold their shape and do not suffer much damage in sunlight. When mixed with each other, or with natural fibres, they can improve the fabric. Many of you sleep in sheets that are made from a mixture of polyester and cotton. These are harder wearing than just plain cotton. They also wash more easily and do not need careful ironing – an important fact when you think how big sheets are.

Activity: Making nylon*

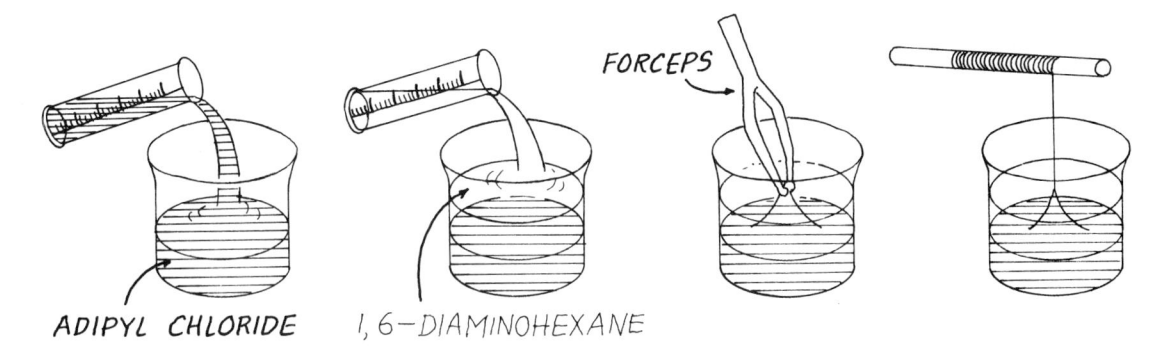

ADIPYL CHLORIDE   1,6-DIAMINOHEXANE   FORCEPS

Here is another experiment that you may be allowed to do at school. You must be careful as the chemicals used are harmful. **If you splash yourself with them you must wash at once with soap and water and then tell your teacher.**

Measure 5 cm³ of adipyl chloride solution into a measuring cylinder and then pour it into a beaker. Wash out the cylinder with lots of water. Measure 5 cm³ of 1,6-diaminohexane solution into the cylinder. Pour this slowly onto the adipyl chloride. The two liquids do not mix.

Dip your forceps in where the liquids meet. Pull up quickly but smoothly as though you had got hold of something. You will find that you have! It is a fibre of nylon. Attach this to a glass rod and then wind smoothly and evenly. At the end you should have a dry nylon fibre wound round the rod.

Does this fibre stretch? Do you think it is thin enough to use for clothes?

You can also get some nylon thread and test it in the same way as the other fibres.

Questions

1. Why are man-made fibres called synthetic?
2. Why was nylon given its name?
3. Make a list of all the useful features of the synthetic fibres.
4. For each of the following brand names, say what fibres they come from: Banlon; Celon; Courtelle; Crimplene; Dacron; Miralon; Perlon; Trevira.
5. Copy and complete these sentences: Raw materials are mixed to form a _____ _____ liquid. This is _____ through fine holes. This makes _____. These are dried in _____. The threads are then wound on to _____. Before they are used the fibres are _____.
6. Why do you think nylon is made into staple yarns?

Unit 10 Weaving

Having obtained the sort of fibre you want it has to be made into a fabric. The main way of doing this is by weaving. Weaving is a very old trade and has gone on in some form for centuries.

The idea behind weaving is simple. It is a way of connecting fibres so that they interlock. A frame is set up and fibres are stretched along it. These are called the **warp fibres**. Another thread is then moved in and out between the warp fibres. In simple weaving the fibre goes over one warp thread and under the next. At the end of the frame the fibre is turned round and comes back. Where it previously went over it is now put under a thread and so on. The fibres running across the weave are called the **weft threads**. The pattern made by this sort of weaving is called **plain weave**. Different patterns can be made quite easily. Instead of going over one and under the next, you can do it in twos or any other combination you like.

This method of weaving is very slow. Obviously it would be quicker if the weft could be threaded faster. This was done very early in the history of weaving. A frame or **loom** is made. This holds the warp threads. The warp threads run through a holder called a **heald**. There are at least two of these on the loom and they hold alternate threads. The healds can be raised and lowered separately. When this happens a space is left between the warp threads. The weaver then throws the weft thread through the gap. The weft threads are kept in a **shuttle**. Try taking a ball and throwing it from hand to hand with your arms a long way apart. The weaver had to be very skilful to catch the shuttle each time. Also the width of the cloth made could only be as wide as the weaver could reach. The weft threads had to be patted into place using a piece of wood called a **reed**.

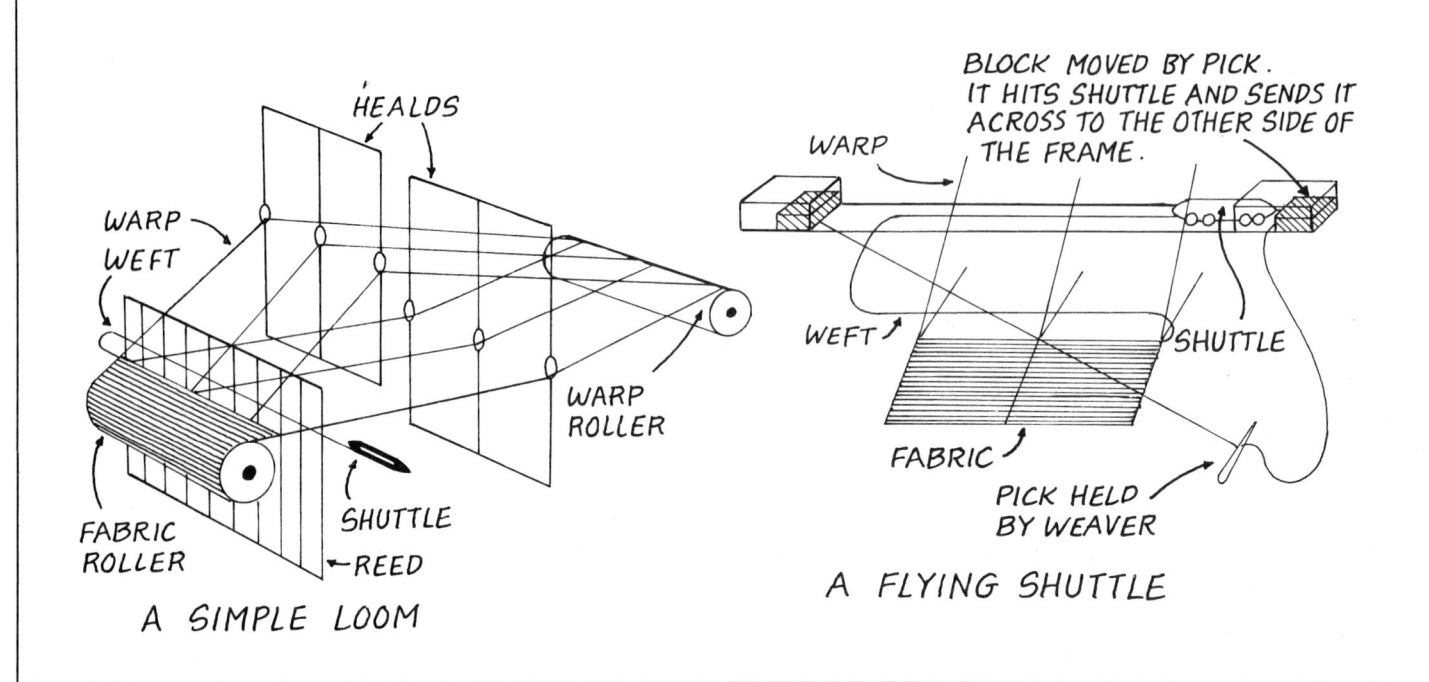

A SIMPLE LOOM

A FLYING SHUTTLE

In the eighteenth century John Kay invented a **flying shuttle**. This was a shuttle that the weaver could work with one hand by a system of pulleys. This left his other hand free to work the reed. This doubled the speed of weaving. It also meant that much wider pieces of cloth could be made.

Later the loom became mechanized. The shuttle, treadles (which move the healds) and reeds were operated by power. Even with power looms a skilled weaver is still needed to set the warp up and to make sure that the weft threads are properly interwoven. The change over from hand weavers to factories making cloth brought much hardship. Women and children often worked from 5 a.m. to 7 p.m. with only a half-hour break for lunch. The children could be as young as six or seven. Some workers rioted to stop the machines. But eventually the machines came and later conditions improved for the workers.

Activity: Making a simple loom

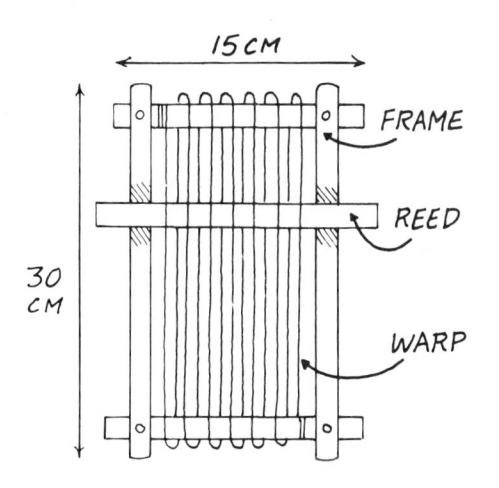

To help you to understand what weaving is all about try making a simple hand loom. Get a metre of wood, 2.5 cm × 3 mm in cross section. Cut this into four pieces; two should be 30 cm long and the other two 15 cm long. Tie, glue or nail them together to make a frame as shown. Get some knitting wool and wind this around the short arms of the frame as shown. Tie them fairly tight. These are the warp threads. Get a thin strip of wood to act as the reed and thread it over and under the warp threads. Take some more wool and thread it into a bodkin. Wind all the wool around the bodkin. Tie one end to the bottom of the warp fibres on one side. Get weaving! Try not to pull too tightly or the edges will be crooked. Remember to go over in one direction and under in the other. You can weave a doll's scarf for a niece or cousin this way. See how slow it is? Now you begin to see why people invented quicker ways of weaving.

Questions

1. Explain the difference between warp and weft.
2. Look at as many different types of fabric as you can. Put a sample in your book with a drawing alongside of the type of weave. Try to find out what each sort is called.
3. Copy the drawing of the loom. Explain how a loom works.
4. What were the advantages of the flying shuttle?
5. When weaving became mechanized the hand weaver was put out of work. Many songs were written about the problems of the times. One called 'Poverty Knocks' starts 'Up every morning at five, It's a wonder that we're still alive'. See if you can find out more about the weaving trade from songs and history books. What sort of conditions did people work in? What hours did they work?

page 23

Unit 11 Knitting

Woven cloth has been used for over 12 000 years but the first knitting is only known from 2000 years ago. It is thought that the Arabs were the first people to invent this method of turning yarn into cloth. The knowledge was brought into Europe by traders. In England guilds were formed for knitting just as for other crafts. The Elizabethan age has been called the 'Golden Age of Knitting'. By the end of her reign Elizabeth I had over 300 pairs of knitted silk stockings. Most of the people who knitted at this time were men. They made very high quality items, possibly because they were fined heavily for bad work! However, mechanization came early. In 1589 William Lee invented the first knitting machine. Even modern machines are very like the one he designed. This invention led to a drop in hand knitting but it survives in some places as a local craft industry and is also a very popular hobby. In Fair Isle and the Shetlands people still knit sweaters and shawls of high quality. Aran sweaters are still hand knitted. Today, knitting by hand is better value than buying shop knitwear.

Knitting is a way of turning a continuous thread into fabric. The thread is looped and each row of loops hangs from the one above it. This is called **weft knitting**. The rows run across the fabric. By using several needles, or a circular needle, it is possible to weft knit in a circle to make tubes. In **warp knitting** the same idea is true but the rows run up and down the fabric at the end.

Knitted garments are flexible (they give) without creasing. They can be fashioned to shape, be plain or patterned. By changing the stitch size they can be made tight or open. They are warm because there are air spaces between the stitches. They do have some disadvantages. They tend to go out of shape. (Ever seen a baggy jumper?) Also, if one of the loops

Fair Isle Sweater

Lace and crochet

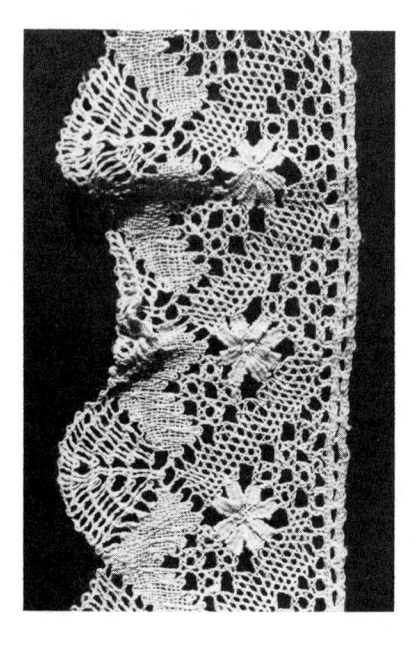

breaks the row hanging from it tends to collapse. This causes a run, or, when it happens in stockings, a ladder.

There are other ways of making fabrics from yarn. For example, **lace work** and **crochet**. They rely on winding the threads around themselves to form an open network fabric. The fabric can be built up on a frame made from pins, or it can be made of loops which are then stitched together.

Finally, it is possible to bond fibres together without weaving, knitting or looping. The fibres can be melted into one another. Such fabrics are not very strong but can be useful for jobs where only a short life is needed, e.g. disposable cloths for cleaning.

Activity: Learning to crochet

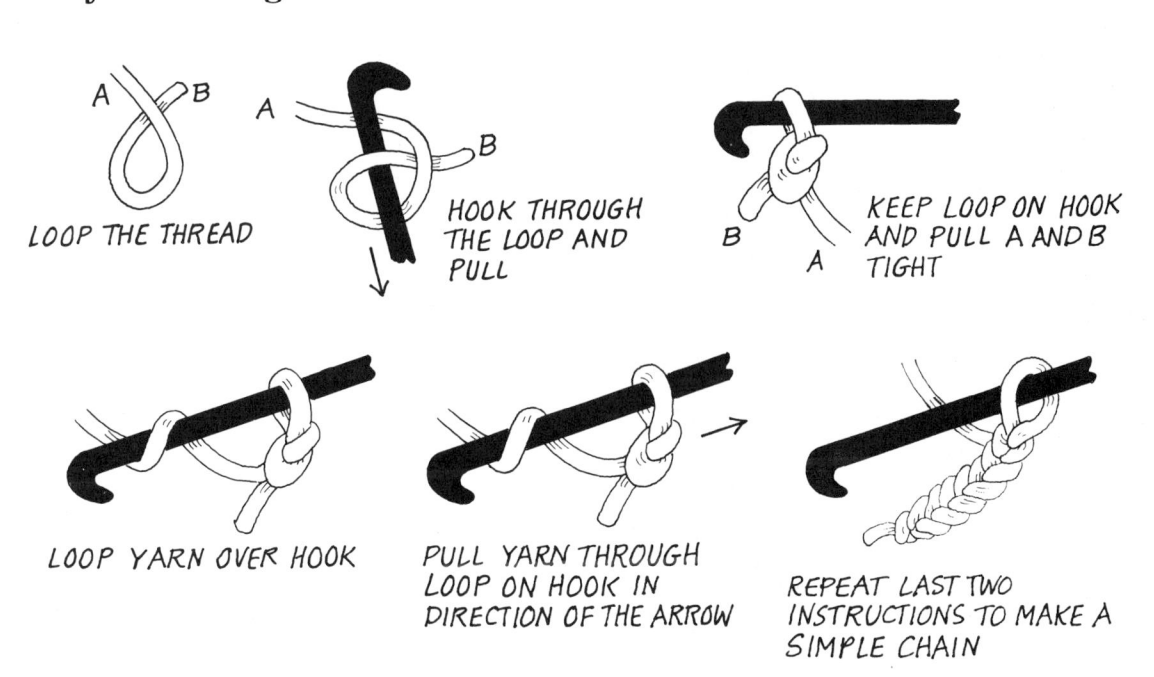

LOOP THE THREAD

HOOK THROUGH THE LOOP AND PULL

KEEP LOOP ON HOOK AND PULL A AND B TIGHT

LOOP YARN OVER HOOK

PULL YARN THROUGH LOOP ON HOOK IN DIRECTION OF THE ARROW

REPEAT LAST TWO INSTRUCTIONS TO MAKE A SIMPLE CHAIN

All you need for this is a crochet hook (size 4.50) and some double knitting yarn. The idea of crochet is to loop the yarn around itself. The instructions above tell you how to make a simple chain. If you are interested go to your local library for further books on how to make proper garments.

Questions

1 Explain the differences between warp and weft knitting.
2 Make a list of the advantages of knitted garments.
3 What are the disadvantages of knitted garments?
4 Describe two other ways of making fabric from yarns. Try to find out if there are more than the ones mentioned in the text.
5 Find out where in Britain knitting is still a craft industry. Mark these places on a map of the British Isles. What do these places have in common?
6 Try to find out the names of some non-woven fabrics.

Unit 12 Finishing

Fibres which have been woven or knitted to form fabrics have to be **finished**. Newly made fabrics are usually dull and unattractive. They come off the loom a grey colour. They often have marks on them from the machines. They feel coarse and have other dirt in them that was in the original fibre. So, before anything else can be done, the fabric must be **cleaned**. This is usually done by washing it in soap and warm water. Sometimes chemicals are added to dissolve the grease. Wool is washed in ammonia but cotton has caustic soda added to clean it. After the cleaning the fabric is **bleached** to whiten it, or it may be **dyed**. Man-made fabrics have a **blue whitener** added as they otherwise look rather yellow. The blue agent fools the eye into thinking the fabric is whiter than it actually is.

Next the fabrics have other finishes added. This is to give the fabric some property it would not normally have. One of the commonest finishes is to **pre-shrink** a fabric. When fibres are woven they get stretched. If you wet them they tend to go back to their original length and so anything made from them shrinks. Natural fibres, when wet, take in water and swell. This also causes them to shrink along their length. With cottons the fabrics are pre-shrunk by steaming them or by adding a resin that stops the fibres taking in water. Woollens are treated to stop the scales sticking together and felting.

Some fibres and fabrics crease badly when worn. These creases tend to drop out of wool, but they stay in cotton. A resin is added to cotton so that the fibres are made more flexible. They do not crease so much. Adding resin does two jobs, it makes cotton **shrink resistant** and **crease resistant**. Resin also helps the fabric to get rid of water because the water no longer soaks in. This makes the fabric **drip-dry**. Man-made fibres tend to drip-dry anyway because they do not soak up water.

Many of us own garments that are **showerproof** or **waterproof**. To be totally waterproof the garments are coated in rubber or oil, or made from plastic.

Flameproof garments are also made. This is particularly important for children's clothes, especially nightwear. A chemical is added to the fabric that stops it burning or melting. It will char but not flame. The fabric may lose some of its softness because of this treatment but is clearly safer. It is worth it to prevent a death by burning.

There are many other finishes. All are used to improve the usefulness of the fabric.

FIBRE AFTER WEAVING (STRETCHED)

SHRINKS AFTER FINISHING

SHRINKS AND SWELLS ON WETTING

Activity: Home-made showerproofing*

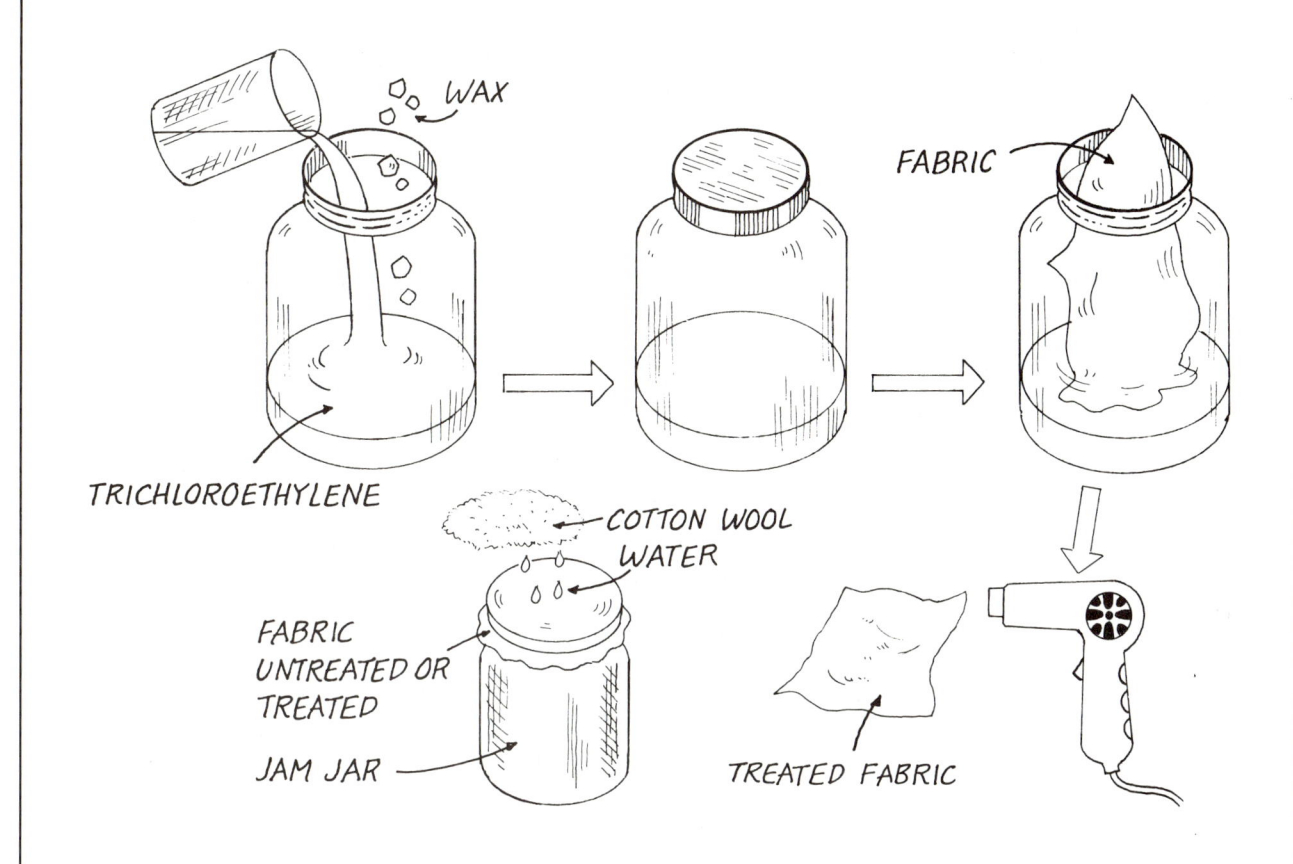

Get some pieces of cloth that have not been treated for showerproofing. It is best if they are made from one fibre only. In a jar with a screw top put 100 cm³ of trichloroethylene liquid. Add to this 5 g of shredded paraffin wax. Screw the top up and leave it until all the wax dissolves (it may take some time). Fix a piece of cloth over the neck of a jam jar using a rubber band. Squeeze some water over it from a lump of soaked cotton wool. Get another piece of cloth and soak it in your wax mixture. Dry it off (a hairdryer is useful). Put this dried piece of cloth over another jam jar and squeeze water onto it. What is the difference? You might like to compare your method of showerproofing with shop products. It is possible to buy aerosol cans for showerproofing. If you use this on untreated cloth you can measure how many drops of water are needed to soak the cloth. Compare this with your own home-made method. Which is best?

Questions

1 Explain why fabrics need to be cleaned when they first come off the looms.

2 How does a blue whitener work? Why is it used on man-made fibres?

3 Make a list of all the things a resin finish does. Explain how each one works.

4 What is preshrinking? How is it done?

5 Below are some trade names for finishes. Try to find out what each one means. What sort of finish does the garment have?
 (a) Sanforized   (d) Pyrovatex (C.P.)
 (b) Dylan   (e) Trubenized
 (c) Mercerized (f) Fixaform

Unit 13 Colour everywhere

Dyes

Tie-dye dress

When you look around you, colour is everywhere. Most mammals only see in black and white but we can see in colour. Colour is important to us in helping us keep out of danger. It is not surprising that humans have always tried to find ways of colouring the things they use. Early on we learnt to paint on cave walls. We painted scenes showing hunting. Maybe they were used in some sort of worship. To do this we learnt to grind up rocks to make bright coloured powders. These could then be used by mixing with water. It did not take long before people started using these chemicals to paint skin and clothes. We call chemicals that stick onto clothing **dyes**.

The ancient British people were called the painted people because they dyed their skins with a chemical called **woad**. This is made from the woad plant, which is related to the cabbage plant. Most people, though, dyed their clothes and not their skins. In ancient times the secrets of how to make a dye could be very important. It could make people rich. The Tyrians made a purple dye which at that time no-one else could make. This meant they could sell their cloth at a much higher price because people wanted this rare colour. We now know that it was made from shellfish. All early dyes were made from animals or plants. These are called **natural dyes**. The ones that work well are very limited and have gradually been replaced by **synthetic dyes**.

Synthetic dyes were discovered by a chemist called W.H. Perkin. In 1856 he was trying to make a drug called quinine. By accident he found that some of the things he had made were very good dyes. The first one he made came from coal tar and was a bright purple colour. Since then dyeing has become a science. Today, most dyes are synthetic. This is because you can get a wider range of colours from them. The colours are brighter than natural dyes (compare old costumes in a museum with modern clothes). It is also easier to control the colour of synthetic dyes. Natural dyes tend to vary in strength and in shade, depending on how fresh the plant or animal is when it is used. Finally, synthetic dyes are cheaper to prepare.

Although there are many different dyes none of them can be used on all fabrics. A dye has to be put onto a fabric in such a way that it does not wash out. It has to cope with sweat, sunlight, detergents and soap, and possibly dry cleaning. No dye is good at doing ALL these things. This means that you have to choose the right one for the job you want. It is no good making curtains from a material whose colour fades in sunlight!

Similarly, different fabrics react in different ways to dyes. Natural fibres take in dyes quite easily because they absorb water. But man-made fibres are hard to dye because water (and dye solutions) will run off them.

Activity: Making your own dyes

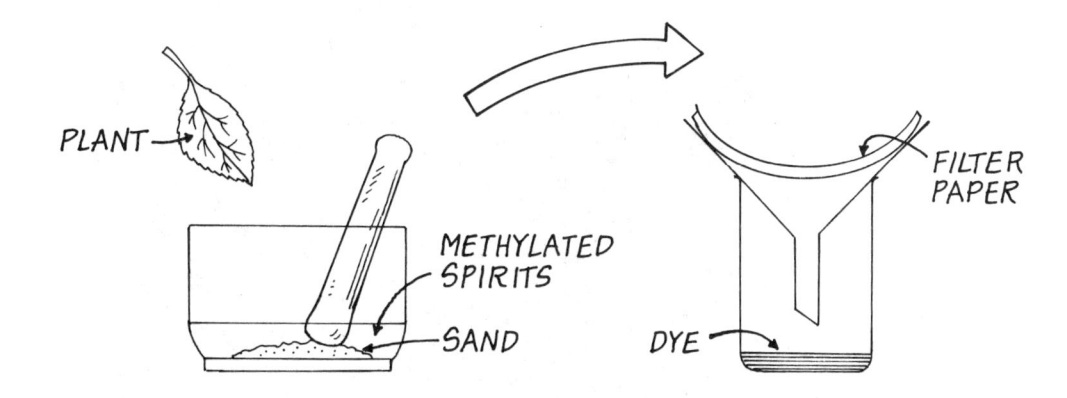

BE CAREFUL not to get the dye all over you or your clothes! Wear an apron or overall and use gloves.

Many different plants can be used to make dyes. Chop up the plant you are using and grind it using a pestle and mortar with a little sand and methylated spirits. When the methylated spirits goes a dark colour, filter it into a clean jar. Try using a lot of different plants. When you have got several dyes, get some clean pieces of old cotton and nylon cloth. Old white sheets are best for this. Cut the sheets into even sized pieces. Dye each piece by leaving it in the dye extract for a minute. Take the pieces out and dry them in an oven, or with a hair dryer. At the end of this process you can make a chart showing what colour dyes you made from the plants and whether they dyed nylon and cotton differently.

Try some of these plants:
gorse; onion skin; nettle leaves; bracken; blackberry; dock root; sloe; elder; turmeric.

Questions

1 Make a list of ways in which colour is important to people.

2 Copy and complete these sentences: The ancient British people were called the _____ people. This is because they dyed their skins blue with _____. This colour can be got from a plant related to a _____. People tried to keep their dye recipes a secret. This helped them get _____ as people had to pay whatever they were asked if they wanted the coloured cloth.

3 Make a list of the reasons why synthetic dyes have replaced natural ones.

4 Write a few sentences about the conditions that dyes might have to stand up to once they are on a fabric.

5 Why are natural fibres easier to dye than synthetic ones?

6 Find out the names of some dyes that you can buy for use at home.

PLANT	COLOUR MADE	HOW DOES IT DYE COTTON?	HOW DOES IT DYE NYLON?

Unit 14 Dyeing

Fibre loaded into vat for dyeing

No dye can be used on all fabrics. This is because each has its own special way of holding onto the fabric. Some fabrics, like wool, are easy to dye. They are made from proteins and react very easily with many other chemicals. This means that quite a lot of different dyes can react with the fibres and join onto them. Some dyes stick to the fabric as though they had dissolved in it. Yet others have to be dissolved in alkalis and then joined to the fibres because they do not dissolve in water. Some new dyes react with the fibres to make a completely new substance altogether. However it is done, the idea is to make the dye stick to the fibre and stay there.

Some dyes have to be helped to stick. The cloth is first treated with a **mordant**. This comes from a Latin word meaning 'to bite'. The mordant holds onto the fibres and the dye holds onto the mordant. By using different mordants it is possible to get different colours from the same dye.

Dyeing can happen at any stage in the manufacture of cloth. It is possible to dye the fibres before spinning, the yarn after spinning, or the finished fabric. The different methods all have their advantages and disadvantages. If you dye the raw fibres there is more chance that they will be evenly coloured. None of the fibres will stop the dye getting to other fibres. Different coloured fibres can then be spun together to give a multi-coloured yarn. It also means that if the different batches of dyed fibres are slightly different shades you can mix them in when spinning and it will not show up. The main disadvantage of this method is that you have to guess which colours will be fashionable well in advance. It may be a year before the fabric made from the yarn comes onto the market. If the manufacturer chooses the wrong colour he will lose a

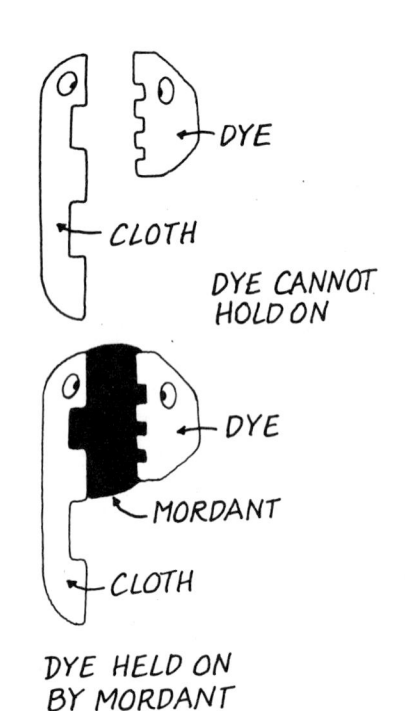

DYE CANNOT HOLD ON

DYE HELD ON BY MORDANT

lot of money. The same problems and advantages are also true for dyeing the yarn.

If you dye the fabric there is no problem of being caught with an unfashionable colour. There are some problems though. It is difficult to make sure that the fabric is evenly dyed, especially if it is thick. The dye may not soak through to the middle properly. This can lead to pale areas appearing as the fabric wears. If the fabric wears unevenly this can make the garment unsightly. Also, since the fabric will be dyed in different batches, the shade may not be even in all of them. This can be difficult for a dressmaker who wants to make a lot of identical clothes.

Activity: Permanent dyeing*

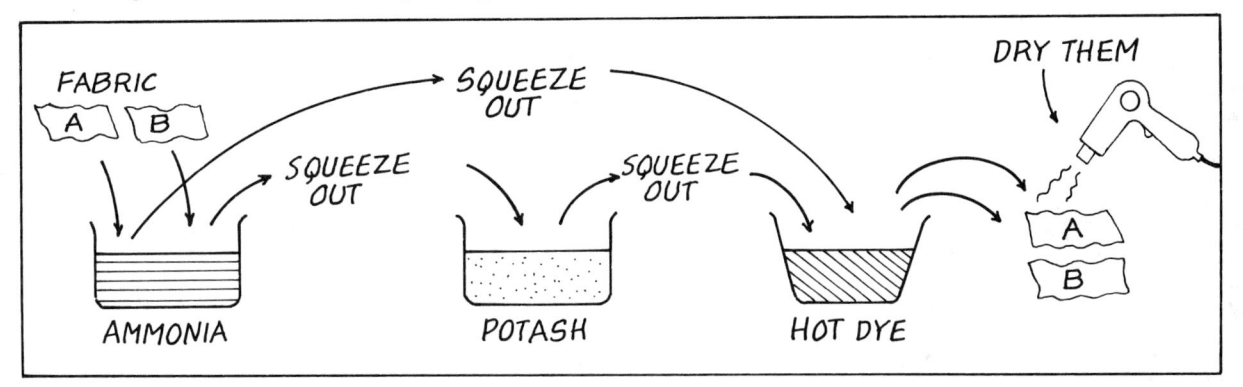

You will have found that not all the dyes you used in the last activity were equally good. Some do not stain nylon very well and some of them fade very quickly. Others wash out easily. It is sometimes helpful to use a mordant. You will need some ammonia solution, some potash alum and some alizarin dye.

Take two pieces of old, clean cotton cloth. Dip one piece of cloth in the ammonia solution and then squeeze it out. Drop it into hot dye (DO NOT BOIL). Take the other piece of cloth and dip it into the ammonia. Then dip it into the potash alum (this is the mordant). Squeeze it out and put it into the hot dye. Leave the pieces of cloth in the dye for ten minutes. Squeeze them out and dry them. When they are dry you can cut them up and test them in the following ways: wash one piece of each in detergent; dip a piece of each into methylated spirits; dip a piece of each into bleach; dip a piece of each into vinegar. Which one keeps its colour best? Has the mordant helped make a permanent dye? Try leaving the pieces in sunlight for a few weeks. Does the colour fade?

Questions

1 Copy and complete these sentences: Dyes hold onto fibres in different ways. Some _____ with the fibre after reacting with it. Others spread through the fabric as though they had _____. Some have to be dissolved in _____ before they will stick to fibres because they do not dissolve in _____. Some new dyes make _____ substances when they react with fibres.

2 What is a mordant and what does it do?

3 Explain what the word 'dyeing' means.

4 Find out how to 'tie and dye' a piece of fabric.

Unit 15 Printing a pattern

Another way to add colour to fabrics is to print the dye on. Printing can be done in several ways. They all use the same dyes that we have already mentioned. In printing the dyes do not soak through the fabric. Instead they are put onto **one side** of the fabric. This means that printed fabrics have a 'right' and a 'wrong' side. There are lots of ways of putting the pattern onto the surface of the fabric.

Block printing You might have done this on paper when you were younger using potatoes and ink. A block is made and a pattern cut in it. The pattern is raised up from the block. The block is usually made from wood. Dye is put on the raised pattern. This is then pressed onto the cloth, which is held flat. The pattern is repeated by adding more dye to the block and moving it along. This is one of the oldest methods of printing.

Screen printing A screen is made from silk stretched on a frame. Nowadays, we use nylon and glass fibre screens as well as silk. The pattern is marked on the screen. The bits you do not want to colour are blocked out. This is done by painting the screen with a lacquer. The screen is then put on the cloth. The dye is put onto the screen and pressed with a roller. This squeezes it through onto the cloth. The dye only goes through the unlacquered parts of the screen. The screen is lifted and the cloth moved along so that the next part can be printed. If you want to make a pattern with more than one colour you have to make a screen for **each** colour, blocking out **all** the other coloured areas. It is very important to get the screens lined up if you use more than one. This method is used for large designs, or those with very fine colour mixing. However, it takes a long time to make screens and so can be expensive.

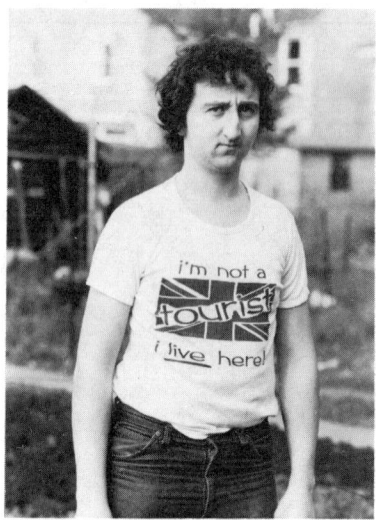

Printed teeshirt

Engraving a copper roller

Roller printing This is a kind of advanced block printing. The design is put on a roller. The cloth is rolled over the roller which has been coated with dye. This process can be done by machines. It is quick and cheap and a regular pattern is made. The differences between this and block printing are first that the pattern is engraved. This means it is cut **into** the roller. The dye sits in the hollows of the pattern. The cloth has to be squeezed hard against the pattern to make sure that the dye gets to it. Secondly, the pattern can be put on the cloth without moving the roller out. This means that the pattern will be regular. Each colour in the pattern needs its own roller. The rollers have to be fitted carefully so that the pattern fits together.

Activity: Batik

Try this method of dyeing first on old clean cloth until you get used to it. Pin the cloth over the open end of a cardboard box. Melt some wax. Do this by putting the wax into a beaker. Put the beaker into a larger container of hot water. This way the wax cannot catch fire. Using a paint brush, paint the design you want on the cloth in wax. Make sure that the wax soaks into the cloth. Make sure that the cloth is stretched tight. When the wax has set, dye the cloth. Do this using a cold water dye. The dye should be in a shallow dish that the whole cloth will fit into. If you have to fold the cloth, the wax will crack and leak. When the dyeing is finished dry the cloth. If you want to make a two-colour design use a pale dye first. Then put the cloth back on the box. Paint on more wax and re-dye in a darker colour. At the end of all this wash the cloth in hot water. This will melt the wax out of the cloth. Then wash it in detergent to get rid of the last tiny pieces of wax. Be careful to rescue as much wax as possible. It can be used again and again. Make sure that you do not let the wax go down the sink where it will harden and block the sink.

Waxing cloth for batik

Questions

1. What is the difference between printing and dyeing?
2. Explain in a sentence how you would know the 'right' side of a printed fabric.
3. (a) What is block printing?
   (b) What are the disadvantages of this method of printing?
4. What kinds of pattern is screen printing used for?
5. Describe in a few sentences how screen printing works.
6. What is an engraved pattern?
7. What are the advantages of roller printing?

Unit 16 Keeping fabrics clean

As you will know, fabrics very easily get dirty. Clothes particularly get a lot of dirt from all sorts of places. Dust and dirt gets trapped on them. Soot and other chemicals from fires collect on them. Grease from sweat sticks to clothes. Food can be dropped on them . . . the list goes on and on. This sort of dirt collects on the fibres and rubs them. Some of the chemicals attack the fibres and dissolve them. If the dirt is left in the clothes they will wear out a lot quicker.

The earliest way of cleaning clothes was simply to soak them in water. Water is very good at dissolving things. If the dirt dissolves and the clothes are shaken, the dirt will be washed away. People used to do this in rivers, beating the clothes against stones. The trouble with this is that water will not clean off grease. Also, all the rubbing and banging wears the clothes out. People soon found they could add other chemicals to help get rid of dirt. **Soap** was a very early discovery but until the 1900s it was a luxury.

Soap is a chemical that makes water spread out and soak into fabrics more easily. This means that the water really gets into the fibres. Soap also joins with and surrounds dirt and grease. Because the soap surrounds dirt, the dirt lets go of the fibres. If the cloth is then shaken the soap and grease are left in the water. The cloth is then rinsed to get rid of any extra soap. It is now clean. Using soap not only gets rid of all sorts of dirt it reduces the amount of shaking and rubbing needed. This means clothes last a lot longer.

Soapless detergents have also been made recently. These do the same job as soap. They have the advantages that they can be made quickly and cheaply. Also a lot of extras can be added to them to help your wash, for instance, a 'blue whitener'. Detergents of this sort are also useful in **hard water areas**.

How scum is formed

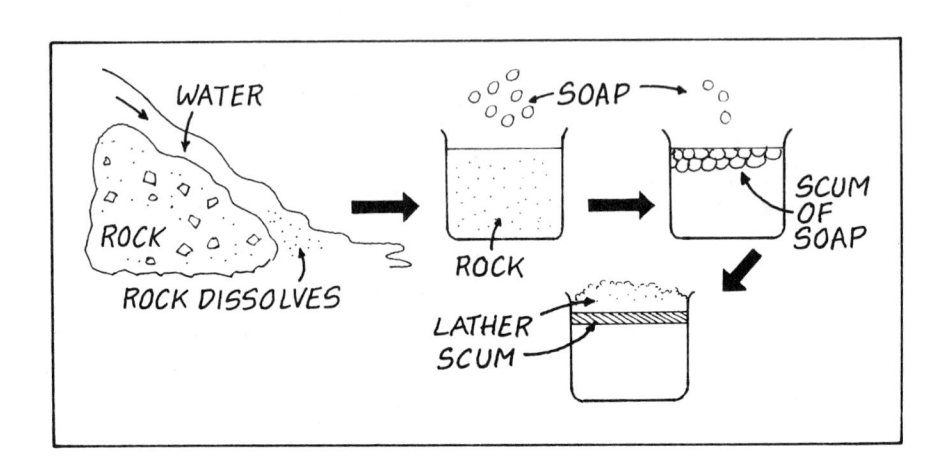

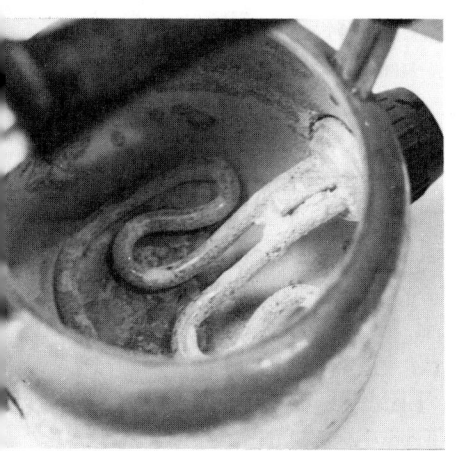

When water soaks through certain rocks some of the chemicals dissolve. This makes the water hard. When you add soap to hard water some of the soap is used up when it joins with the dissolved chemicals. This happens before a lather can form. The mixture of soap and dissolved chemicals makes a **scum**. Scum can damage clothes. Soapless detergents get rid of the dissolved chemicals without making a scum.

The effect of hard water on a kettle

Activity: Making soap*

Preparing a simple soap is quite easy. (Be careful when you are heating and take care with caustic soda. Wear safety glasses.)

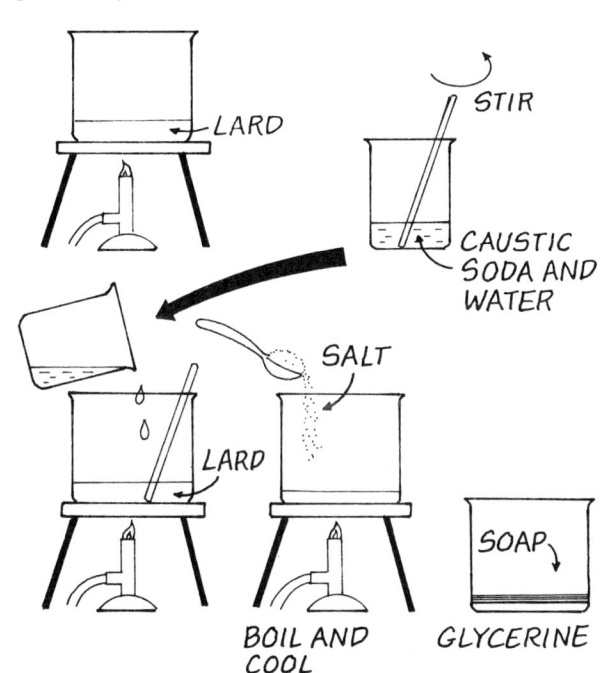

Weigh out 25 g of lard and melt it in a beaker. Use a VERY SMALL bunsen flame to do this or you will find the fat boils. Put 100 cm^3 of water into another beaker. Add to this 5 g of caustic soda and stir it until it dissolves. When the fat has melted, carefully add the solution of caustic soda. Stir the mixture well and then heat it on a low flame for about an hour. As it cooks the mixture will go thick. When it is thick and has been reduced by about a third add a teaspoonful of salt. Boil this new mixture for five minutes and then let it cool. When it has cooled you will find two layers. The top layer is soap. Take the soap out of the beaker. What is left behind? Does it remind you of anything? Glycerine is made from this.

Test your soap by seeing how good it is at lathering compared to a proper shop soap. You might try washing two similar dirty bits of cloth. Use your soap on one and a shop soap on the other.

Questions

1. Make a list of all the different ways in which clothes can get dirty. Think up ones of your own that are not mentioned.

2. How does dirt damage clothes?

3. Explain, with diagrams, how soap removes dirt from clothes.

4. What advantages do soapless detergents have over soaps?

5. What is hard water?

6. How can hard water damage clothes if you wash them in soap?

7. Try to find out how soap and soapless detergents are made in industry. Write a few sentences about each process.

Unit 17 Washday problems

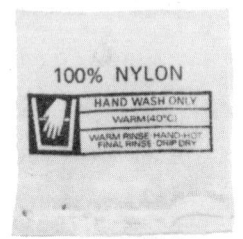

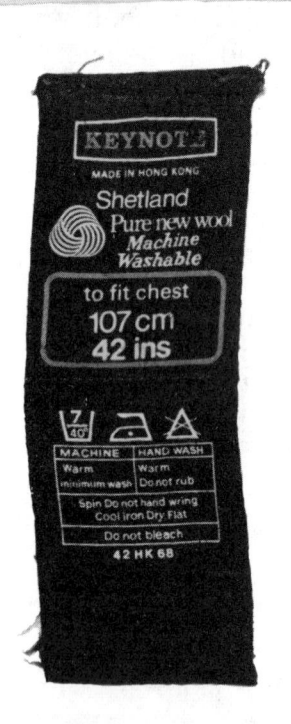

You have seen that washing things in soaps and detergents keeps them clean. It is also true that hot water will work better than cold. This is because more dirt dissolves in hot than in cold water. But not all fabrics will stand up to the same treatment. For instance, you already know that wool will 'felt' if you rub it too hard. Some fabrics will shrink in hot water but not in cooler water. When there were only a few sorts of fabric it was easy to know how to wash them. Since man-made fibres were invented, though, things have become harder. How are we supposed to remember all the different ways to wash each fabric? This problem led to a special labelling system in clothes. The idea of these labels is to tell you exactly how to look after your clothes. They state the temperature that the clothes should be washed at. They tell you about the method of washing, how to dry the clothes and how they can be ironed. These labels should be sewn into the clothes you buy so that you can always find out how to clean them.

The people who first thought up the system decided that the labels should use symbols and short phrases to explain what to do. Four special symbols were invented. These deal with washing, ironing, bleaching and dry-cleaning. As well as this, nine different ways of washing are symbolized on the labels. Each way involves different water temperatures and drying methods. You should always check carefully and wash things with the same labels together. If you ignore instructions you may find clothes shrink. Some colours may run and stain other clothes. Some of the usual symbols are given below.

Symbol	MACHINE	HAND WASH
4/50 tub	Hand-hot medium wash	Hand-hot
	Cold rinse. Short spin or drip-dry	
Crossed triangle	DO NOT USE CHLORINE BLEACH	
Iron	WARM	
P in circle	DRY CLEANABLE	
Circle in square	TUMBLE DRYING BENEFICIAL	

page 36

Activity: Washing and creasing

Obtain a piece of coloured cotton cloth and a piece of woven wool cloth of the same colour. Cut each piece of cloth into eight even-sized pieces. Follow the instructions for each of the first eight washing methods and wash one of each sort of cloth in a beaker by each of the eight methods. Make sure you use the same amount of soap for each wash. When the pieces of cloth have dried compare them. What differences do the wrong washing methods make?

Get some metal strips and wrap a piece of material round each as tightly as you can. Use different materials. Hold them in place with rubber bands. Put each into boiling water in a beaker for five minutes. Unwind the material and hang it up to dry. Which sorts of material crease easily? Do the creases drop out of any of them if you leave them? Try ironing them with a cool iron. If this does not get rid of the creases use a hotter iron. Be careful not to melt any of the fabrics.

Copy the table shown below and fill in your results.

Method	Has it shrunk?	Has it faded?	Has it creased?
1 cotton			
2 wool			

Questions

1 Why do we need labels in the garments we wear today?

2 Copy the different symbols for washing instructions into your books. Explain what each one means.

3 Do you think the symbols used are clear? If not why not? Try to invent some symbols that would be as good or better than the ones in use. Remember that they need to be understood quickly and easily.

Unit 18 **Stains**

Even with good care any garment can get stained. A stain is any mark that sticks to the fabric. It holds on rather like a dye. Stains have many causes. Probably the most common stains come from food and drink and blood from cuts. Other stains include grease and oil from mending cars by the roadside, or sitting on a beach; also grass stains from a cricket match, or sitting on the hillside.

There are three main ways of removing stains. They all use chemicals. Some chemicals dissolve the stain. These are called **solvents**. Other chemicals react with the stain and join onto it. When they are dry they, and the stain, can be rubbed or brushed off. These are called **absorbents**. The third sort are the **bleaches**. These react with the stain to break it up into small bits which let go of the fabric. Many of the chemicals used as stain removers are quite strong. This means you should always wash them out thoroughly after they have been used. If not you may find that they cause more damage than the stain.

Stains are rather like dyes. When they first go on they do not stick very well. The longer you leave them the harder it is to remove them. So the first thing to do is to attack the stain early. The sooner you get to it the more chance you have of getting rid of it. The best thing to do is to soak the stained fabric in cold water as soon as possible. While this will not cure all stains it will help with most common ones. The fabric should be left to soak until it can be dealt with properly.

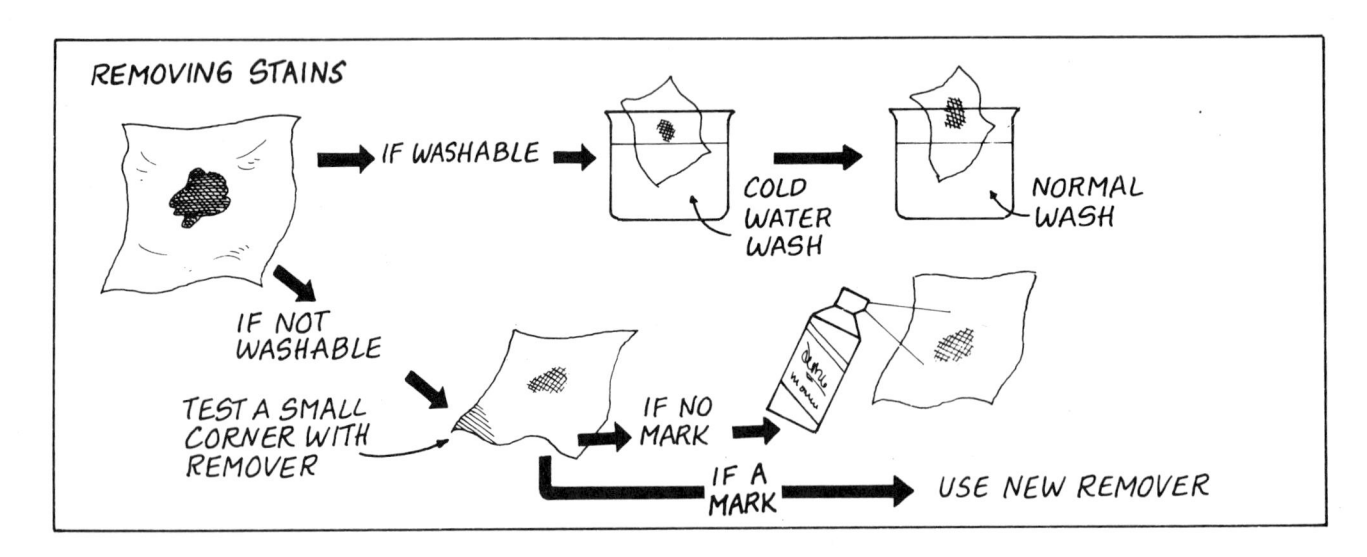

REMOVING STAINS

IF WASHABLE → COLD WATER WASH → NORMAL WASH

IF NOT WASHABLE → TEST A SMALL CORNER WITH REMOVER → IF NO MARK →

IF A MARK → USE NEW REMOVER

Of course, not all fabrics can be treated with the same sort of stain remover. Every fabric has its own special chemistry. If you treat wool with a chlorine bleach you will dissolve it. It is important to know both what the stain is and what the fabric is. This can be a problem today as so many fabrics are mixtures of different chemicals. If you are not sure what the fabric is made from then test a small piece with the stain remover. If the stain remover marks the fabric then choose another remover. In all cases try the simplest stain remover first. Only go on if this does not work.

Most food stains can be washed out, especially if the cloth is soaked as soon as it is stained. Blood stains should be washed in salt and cold water, and then washed normally. Grass and ball-point ink can be dissolved using methylated spirits. Again, the clothes should be washed after removing the stains. Grease spots can be removed using aerosol sprays. These should all be used with great care as the fumes can be poisonous.

Activity: Cleaning stains*

Get several pieces of cotton cloth, all the same size. Stain one with some jam by rubbing it on the cloth. Stain another with some coffee. Stain a third by rubbing it with grass. Stain a fourth piece with some blood (from meat – not your own!). Let all these stains dry. Cut each stained piece of cloth in half. Now wash one half in hot soapy water, rinse it and dry it. Soak the other half in cold water and detergent for a couple of hours. Then give it a hot wash. Compare the two halves. Which one is cleanest? Which stains are easiest to get rid of?

Try doing the same experiment with other fabrics that can be washed in hot water. You may like to try different stains, e.g. butter or ink.

You can also test other stain removing methods. Get some bleach, methylated spirits, salt, an aerosol dry-cleaner and hydrogen peroxide. Get a range of different fabrics. Cut each fabric into five pieces. Put some of the different stain removers on each piece. Make a note of what happens. Can you decide from this simple test which cleaners to use with the different fabrics? Which one should never be used on wool?

Questions

1 Copy and complete these sentences:
   (a) A solvent is _____.
   (b) A bleach is _____.
   (c) An absorbent is _____.

2 Make a list of the commonest sorts of stain.

3 Why do we need to use different sorts of stain remover?

4 What simple method would you use to remove a food stain?

5 What should you do if you have stained a jumper and do not know what it is made from?

6 Try to find out why 'dry-cleaning' is called dry.

FABRIC	WHAT THE CLEANERS DO				
	BLEACH	SALT	METHYLATED SPIRITS	AEROSOL	HYDROGEN PEROXIDE

Unit 19 Choosing the right fabric

These days everyone will have to choose fabrics at some time in their lives. You will obviously buy clothes. When you get a home you will need to choose carpets, curtains and armchairs. You and your family will need sheets and blankets. Understanding the good and bad things about different fibres and fabrics will help you choose sensibly. What then do you need to consider?

First of all you need to consider what the cost will be. Fashion, and what you would like, are all very well, but if you have no money you will be stuck! So you must ask yourself 'how much can I afford?'. Once you have decided this you need only look at the things that come into that price range. The next question you must ask is 'what is it going to be used for?' 'what kind of a job has the material to do?' A shirt for the evenings will not need to be as strong as one which you will use for working in. A chair cover has got to be good at staying in one piece when rubbed a lot. So you look at the things in your price range and ask yourself 'will I get value for money?'

A variety of fabrics in domestic use

The commonest thing you will need to decide is the material for clothes. Below you will find the main properties of different fabrics.

Cotton This is very hard wearing. It takes in water and so is often more comfortable to wear. It is often used in mixtures to give strength. On its own it tends to get dirty quickly and does not iron easily.

Wool This is warm and comfortable material. The air trapped in the fibres keeps you insulated and it lets sweat through. However, it is expensive if used on its own. It shrinks easily and so needs to be treated more carefully when washed. It is often used in mixtures to give added warmth.

Nylon This is tough and drip-dry. It can be washed and hung up overnight and be ready for wear the next day without ironing. It is elastic and can also be given permanent creases. It can make static that causes it to pick up dirt. It can also be uncomfortable in hot weather because it does not take in water. It gradually loses its colour and can fade easily in sunlight.

Polyesters These are lightweight and strong. They are drip-dry and can be given permanent creases. They keep their shape easily. Again, they do not take in much water so they may be less comfortable on hot days.

Obviously mixtures of some or all of these fibres will make fabrics of slightly different properties. Which one you buy will depend on your own needs. If you are unsure consult the manufacturer, shop assistant, or go to the library. A little time can save a lot of money in the long run!

Activity: Identifying fabric fibres*

As you have seen it can be important to find out what fabrics are made of. Suppose you stain your clothes. Suppose you want to make something and have some material. If you do not know what the fibres in the fabric are you will not be sure what to do. What stain remover should you use? Is the material the right sort for the kind of clothes you need? It is possible to work out what the material is made from by doing some burning tests. You should only use very small amounts of the fabrics. The fumes from some fabrics are poisonous. If you are testing something that is already made, take samples from somewhere where it will not show! Follow the fibre finder guide (left) to work out what your fabric contains.

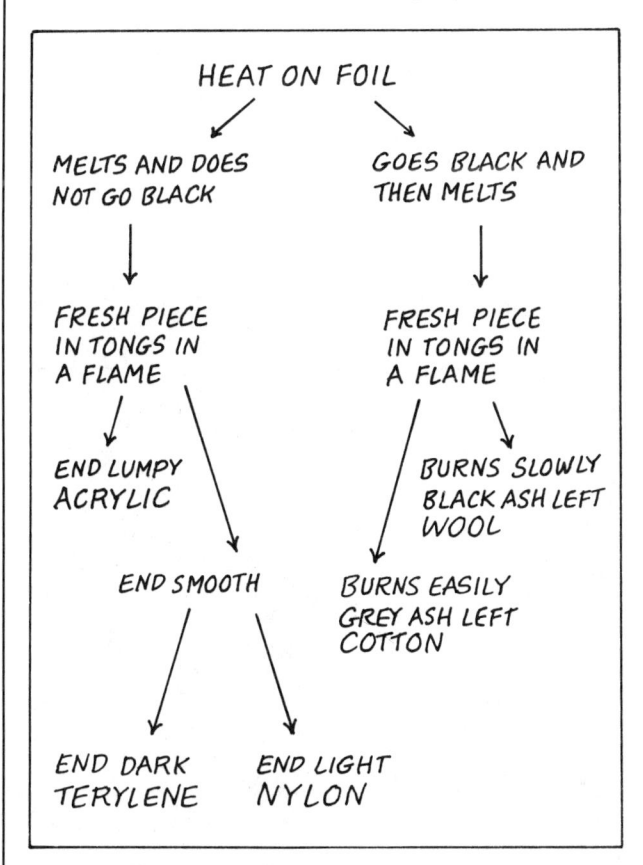

```
                    HEAT ON FOIL
                   ↙            ↘
         MELTS AND DOES      GOES BLACK AND
         NOT GO BLACK        THEN MELTS
              ↓                   ↓
         FRESH PIECE         FRESH PIECE
         IN TONGS IN         IN TONGS IN
         A FLAME             A FLAME
            ↓                 ↙       ↘
         END LUMPY       BURNS SLOWLY
         ACRYLIC         BLACK ASH LEFT
              ↓          WOOL
         END SMOOTH      BURNS EASILY
           ↙    ↘        GREY ASH LEFT
      END DARK  END LIGHT COTTON
      TERYLENE  NYLON
```

Questions

1 Why is it useful to know something about fibres and fabrics?
2 What three things do you need to ask before choosing a fabric?
3 Explain why you need to ask each of these three questions.
4 Using the information given in this chapter and the rest of the book, decide what fibres you would use for the following clothes. Give reasons for your answers. (a) A summer shirt, (b) a winter suit, (c) tennis shorts, (d) underwear, (e) a party dress, (f) socks. If more than one sort of fibre will do say so and explain why. If you want to use a mixture say so and explain why.

Unit 20 Strange fibres

Asbestos rock

Fibres come from many different places. One of the most unusual is **asbestos**. This is a fibre made from a rock. The rock is found as thin veins and is as hard as granite. The rock contains several elements. Calcium, magnesium, oxygen and silicon are all joined together in long chains. These chains are the fibres of asbestos. They can be up to eight centimetres long. Asbestos has been used for a long time and was thought to be magical because it does not burn. The story is told of the Emperor Charlemagne dropping some asbestos cloth into a fire and bringing it out unchanged.

The rock is blasted to get it out of the ground. Then it is crushed to release the fibres. Once the fibres are free and cleaned they can be woven into cloth in much the same way as any other fibres. Asbestos has been used widely for fire fighting. Blankets and clothes can be made from it for firemen. Theatre safety curtains, ceiling tiles and ironing board mats are also made using asbestos. We now know that asbestos can be very dangerous to health. The very small fibres get into the lungs when we breathe. They gradually damage the cells and cause a disease called **asbestosis.** This may not appear for many years after breathing in the fibres. However, because of this most asbestos has been replaced and it is no longer widely used.

In place of asbestos, **glass fibre** is sometimes used. Glass is a very strong material which can insulate and which does not burn. Glass fibre is used to insulate lofts, wrap underground pipes and protect aeroplane exhaust pipes. It has even been made into clothes. If you ever use glass fibre you must take care that small fragments of glass do not go into your hands or lungs. If you insulate your loft with it you should wear protective clothes. One advantage glass fibre cloth has is that colours on it do not fade, so it is very good for curtaining.

Protective clothing in use during loft insulation

page 42

Finally, the last use of fibres that we shall mention is in paper making. The whole of this book is made from fibres of plants pressed together to make thin sheets of paper.

The fibres can be taken directly from the plant or from cloth made from plants (e.g. cotton or linen). Early paper was made from rags. These were shredded and pulped with an alkali (a chemical that is the opposite of an acid). This frees the fibres. After washing and pressing the fibres can be made into thin sheets. On drying they are what we call paper. If you wet a sheet of paper it tears easily. This is because the fibres tend to fall apart. When dry the paper is very strong. Many special methods are used to make paper smoother, and different colours can also be made.

Activity: Modelling with paper

It is quite easy to model with strange fibres. Glass fibre is often used to make models and to repair car bodies, etc. You can make models easily using paper pulp. Tear some old newspaper up and soak it in cold water in a bucket. Leave it there for several hours. After this shred the paper more while it is still in the bucket. Work it with your hands and then bring it out and squeeze it hard. Put the wet (but not soaking) pulp in another bucket. Mix up some ordinary cellulose paste. Be careful to follow the instructions on the packet. Add the paste to the paper pulp and mix well using your hands. You should end up with a mixture which is not too wet and not too dry.

This mixture can be used to make models. Work on an old piece of hardboard or something similar. Use the mixture fresh. Make sure that the bulk of the model is at the base or it will collapse when drying. Try not to make any part too thin as it may break off in the drying. The material can be sculptured quite finely. Leave the model in a warm dry room for two or three days to dry. After this you can paint it with water-based paints or inks. A final layer of polyurethane varnish will keep it looking nice and keep it dry.

Questions

1 What is asbestos and what is it made from?

2 Why is asbestos important?

3 Explain carefully why asbestos is being replaced by other substances.

4 As well as those mentioned in the chapter try to make a list of as many uses of glass fibre as you can.

5 Copy and complete the following sentences:
Paper can be made from old _____.
These are mixed with an _____ to free the _____. The pulp is then _____ and _____ to make sheets of paper. Paper is not strong when wet because the fibres tend to _____.

6 You might like to try to find out how to model in glass fibre. There are many books to help. Remember always to follow the safety rules.

Index

acrylics, 20
America, 14–15, 20
Arabs, 24
Arkwright, Richard, 6
artificial silk, 18
asbestos, 42

Babylonia, 10
batik printing, 33
bleaching, 26, 38
blue whitener, 26, 35
bobbins, 6–7
bolls, 14–15
Bombyx mori, 16
bonded fibres, 25

carbonizing wool, 12
carding wool, 12
Carothers, Dr Wallace, 20
cellulose, 4, 18–19
Chardonnet, Count, 18
Charles II, 10
cleaning, 34–9
cloth *see* fabric
cloth trade, 10
clothes, 4, 11, 15, 17, 18, 20, 28, 34, 36, 38, 40
cocoons, 5, 16–17
combing wool, 12, 14
cotton, 4, 8–9, 14–15, 21, 26, 29, 31, 37, 39, 40
cotton gin, 14
creasing, 12, 26, 37
crochet, 25
Crompton, Samuel, 6

degumming silk, 17
deniers, 8
detergents, 35
drawing cotton, 14
dry-cleaning, 38–9
dyeing, 26, 28–33

Edward III, 10
Egypt, 14
Elizabeth I, 10, 24

fabrics, 4, 13, 15, 17–19, 21–3, 26–7, 30–1, 40–1
Fair Isle, 24
felt, 13, 26, 36

fibres, 4–6, 8–9, 12–13, 15, 18–21, 25, 26, 30, 36, 41
filament yarns, 8, 16, 19, 20
flax, 4–5
fleeces, 12–13
Flemish weavers, 10
fliers, in spinning, 6
flying shuttles, 23

glass fibre, 32, 42
guilds, 10, 24
gun cotton, 18

Hargreaves, James, 6
healds, 22
hemp, 5

Japan, 16
jute, 5

Kay, John, 23
knitting machines, 24
knitwear, 13, 19, 24–5

labelling, for washing, 36
lace work, 25
lapping (cotton), 14
Lee, William, 24
Lincoln, Abraham, 15
linen, 5, 8–9
looms, 22–3

masterpieces, 10
modelling, 43
mohair, 5
mordants, 30–1
muslin, 15

Normans, 10
nylon, 5, 8, 20–1, 29, 31, 32, 41

paper, 19, 43
Perkin, W.H., 28
Peru, 14
Phoenicians, 10
ply, 8
polyesters, 20–1, 41
polyamides, 20
printing, 32–3
pupae (silkworm), 16

rayon, 18–19
reeds, weaving, 22–3
ring frame spinning machines, 6
Romans, 10

Saxons, 10
Schonbein, 18
scouring wool, 12
screen printing, 32
scum, 34–5
sheep, 11–13
Shetlands, 11, 24
showerproofing, 26–7
shrinkage, 26, 36–7
shuttles, 22–3
silk, 5, 8–9, 16–17, 32
silk, artificial, 18
silkworms, 5, 16–17
sisal, 5
slivers, 6, 14
soap, 28, 34–5
spindles, 6–7
spinning, 6–7, 12, 16–17
Spinning Jenny, 6
Spinning Mule, 6
stains, 38–9
staple yarns, 8–9, 15, 20
Swan, Joseph, 18
synthetic fibres, 5, 8, 18–21, 36

threads, 4–7
twists, 8–9
Tyrians, 28

warp knitting, 24
warp threads, 8, 22–3
washing processes, 21, 26, 28, 34–7, 39
water, 26, 34–5; and wool, 12–13
water frame spinning machine, 6
waterproofing, 26
weaving, 8, 22–3
weft knitting, 24
weft threads, 22–3
woad, 28
wool, 4, 10–13, 26, 30, 37, 38, 40
wool trade, 10–11
worsted, 9, 12

yarns, 6–9, 20, 30
Yorkshire, 10